T0280584

Environmental Footprints and Eco-design of Products and Processes

Series editor

Subramanian Senthilkannan Muthu, SgT Group and API,
Hong Kong, Hong Kong

This series aims to broadly cover all the aspects related to environmental assessment of products, development of environmental and ecological indicators and eco-design of various products and processes. Below are the areas fall under the aims and scope of this series, but not limited to: Environmental Life Cycle Assessment; Social Life Cycle Assessment; Organizational and Product Carbon Footprints; Ecological, Energy and Water Footprints; Life cycle costing; Environmental and sustainable indicators; Environmental impact assessment methods and tools; Eco-design (sustainable design) aspects and tools; Biodegradation studies; Recycling; Solid waste management; Environmental and social audits; Green Purchasing and tools; Product environmental footprints; Environmental management standards and regulations; Eco-labels; Green Claims and green washing; Assessment of sustainability aspects.

More information about this series at http://www.springer.com/series/13340

Subramanian Senthilkannan Muthu
Editor

Quantification of Sustainability Indicators in the Food Sector

 Springer

Editor
Subramanian Senthilkannan Muthu
SgT Group and API
Hong Kong, Hong Kong

ISSN 2345-7651   ISSN 2345-766X   (electronic)
Environmental Footprints and Eco-design of Products and Processes
ISBN 978-981-13-4772-6   ISBN 978-981-13-2408-6   (eBook)
https://doi.org/10.1007/978-981-13-2408-6

This Springer imprint is published by the registered company Springer Nature Singapore Pte Ltd.
The registered company address is: 152 Beach Road, #21-01/04 Gateway East, Singapore 189721, Singapore

This book is dedicated to:
The lotus feet of my beloved
Lord Pazhaniandavar
My beloved late Father
My beloved Mother
My beloved Wife Karpagam and
Daughters—Anu and Karthika
My beloved Brother

Contents

Business Strong Sustainability Performance: Evidence from Food Sector

Ioannis E. Nikolaou and Thomas Tsalis

Abstract A composite sustainability index is suggested to evaluate strong sustainability of businesses by utilizing the triple-bottom-line approach and principles of strong sustainability. On the one side, the proposed index is classified into three classical aspects of sustainability to measure financial performance, environment protection, and social justice. On the other side, it focuses on some basic concepts of environmental science in relation to carrying capacity, safe minimum standards, and critical capital mainly to design certain thresholds for indexes which businesses should attain. An application of the proposed methodology has been made to the food industry in order to draw some implications useful to overcome some of the shortcomings of previous studies. The findings show that the idea of integrating economic, social, and environmental thresholds into corporate indicators might be a good basis to evaluate the strong corporate sustainability performance and offer a comprehend signal to stakeholders.

Keywords Corporate sustainability · Strong sustainability performance
Composite sustainability index · Benchmarking/measuring systems

1 Introduction

The measurement of corporate sustainability has gained a great momentum in the last years. Many different methodologies have been suggested to measure corporate sustainability which could be classified into various categories based on measurement units, aspects of sustainability, and single or composite character of sustainability index (Searcy 2011; Goyal et al. 2013). The first category could be classified into three further types. The first type of methodologies put emphasis on corporate sustainability in financial terms (Atkinson 2000). The second type focuses on

I. E. Nikolaou (✉) · T. Tsalis
Department of Environmental Engineering, Democritus University of Thrace,
Xanthi, Greece
e-mail: inikol@env.duth.gr

© Springer Nature Singapore Pte Ltd. 2019
S. S. Muthu (ed.), *Quantification of Sustainability Indicators in the Food Sector*,
Environmental Footprints and Eco-design of Products and Processes,
https://doi.org/10.1007/978-981-13-2408-6_1

1

nonfinancial terms to measure corporate sustainability (Epstein and Roy 2001) and the third type measures corporate sustainability in mixed terms (financial and nonfinancial) such as eco-efficiency and triple-bottom-line approach (Ilinitch et al. 1998; Nikolaou and Evangelinos 2012). Methodologies of the second category focuses on measuring different aspects of sustainability such as environmental performance, corporate social performance, corporate eco-efficiency performance, and triple-bottom-line performance (Wartick and Cochran 1985; Tyteca et al. 2002; Székely and Knirsch 2005; Nikolaou and Matrakoukas 2016). The final category emphasizes on corporate sustainability by utilizing either single indicators or composite indexes (Singh et al. 2007; Baumgartner and Ebner 2010).

Some important weaknesses of these methodologies are standardization and reliability. It is known that a general accepted methodology to measure corporate sustainability is a very complex task (Searcy 2012). Actually, the concept of corporate sustainability consists of various and complex meanings which are very difficult to be measured. Yet, in the case where scholars employ globally accepted guidelines (e.g., GRI, ISO 14031, and SA 8000) to measure corporate sustainability, this difficulty and complexity are being remained (Searcy 2009). The lack of appropriate information is another significant hinder for developing composite indexes, while many difficulties is arisen in the effort to select suitable corporate sustainability indicators since different corporate sustainability definitions lead to different aspects such as environmental sustainability, eco-efficiency, and triple bottom line (Schmidt et al. 2004; Brattebø 2005; Singh et al. 2007; Hubbard 2009).

Most of the previous methodologies pay more attention on measuring weak sustainability of businesses. It is rare to identify composite corporate sustainability indicator which takes into account the appropriate trade-offs among economic, environmental, and human capital (Faucheux and O'Connor 1998). Only a few studies have conducted to measure business strong sustainability by comparing the business sustainability scores with the average score of the sector. The majority of current methodologies are based mainly on the average score of a sector and not on the significant concepts of environmental science like carrying capacity, safe minimum standards, rebound effect, and critical capital (Figge and Hahn 2005; Van Passel et al. 2007).

To overcome some of these weaknesses, this book chapter develops a methodology to measure corporate strong sustainability by defining initially the content for each aspect of corporate strong sustainability. The innovation of this methodology is based on the combination of corporate performance scores with well-defined thresholds which are associated with basic principles of environmental science. Finally, a case study with three businesses in the food sector has been conducted.

This chapter is classified into four further sections. Section 2 analyzes corporate sustainability indicators and strong sustainability definitions. A description of the proposed composite sustainability index for businesses is presented in Sect. 3. A case study has been carried out in Sect. 4. Finally, a discussion, implications, and discussion for corporate strong sustainability have been described.

2 Theoretical Underpinning

Although the concept of corporate sustainability has been considered very significant, nevertheless, there is no consensus among scholars for this concept (Van Marrewijk 2003; Aras and Crowther 2008; Montiel 2008). So far, many scholars consider synonymous corporate sustainability and corporate social responsibility which focus on a triple goal strategy as follows: (a) economic viability, (b) environmental preservation and social justice (Van Marrewijk 2003). This implies that environmental and social strategy improves the profitability of businesses (Salzmann et al. 2005; Schaltegger et al. 2012).

However, a new theory for corporate sustainability is necessary in order to address more environmental and social side of sustainability and less financial side (Dyllick and Hockerts 2002). The majority of current theories for corporate sustainability are more close to business logic and less to sustainability idea (Lozano et al. 2015). Indeed, many theories focus on strategic management of businesses which deals with addressing of stakeholders' needs (stakeholder theory), creating competitive advantage (resource-based theory and knowledge-based theory), achieving social peace (legitimacy theory), and aligning with requirements of institutions (institutional theory) (Nikolaou 2017).

To evaluate corporate sustainability, many methodologies have been suggested according to a single or much components of sustainability (e.g., environmental, social, economic, and triple-bottom-line performance). The methodologies regarding corporate environmental sustainability methodologies have put emphasis only on measuring environmental aspects (e.g., air emissions and global warming), while methodologies of social issues focus on measuring social and ethical aspects (Veleva et al. 2003; Hutchins and Sutherland 2008). Some economic-based methodologies emphasize on identifying association between businesses financial figures and environmental and social strategies (Schaltegger et al. 2012). Some methodologies measure eco-efficiency performance or triple-bottom-line performance. Finally, there are two trends for methodologies in relation to the way in which they design indicators. The former category of methodologies provides sets of various single sustainability indicators and the later category offers composite sustainability indexes.

Some of the shortcomings for methodologies for single indicators are the absence of standardized indicators (Olsthoorn et al. 2001), the use of mixed units (Delai and Takahashi 2011), the various definitions for corporate sustainability (Delai and Takahashi 2011), the focus on particular business sector (Rahdari and Rostamy 2015), and the lack of leading and lacking sustainability indicators (Figge et al. 2002). The shortcomings of composite index methodologies are summarized as the unclear techniques to integrate single indicators as well as the lack of clear techniques to normalize the measurement units (Singh et al. 2007). Composite indexes have also high measurement complexity and great level of information losses in the procedures of integration (Salvati and Zitti 2009; Sridhar and Jones 2013). Subjectivity exists also on the procedures of evaluating weight factors which are required to incorporate single indicators into the final composite index (Shwartz et al. 2009). Moreover, the majority of suggested composite indexes are not able to incorporate the concept substitution among three types of capital (e.g., economic, environmental, and human resources capital).

The substitution of three types of capital classifies businesses into two categories. In the first category, businesses contribute to week sustainability. The majority of relative literature and theories focuses on weak sustainability. Initially, stakeholder theory determines business sustainability with respect to the needs of stakeholders without taking account specific allocations among three types of capitals (economic, environmental and social) in order to be achieved the goals of strong sustainability (Steurer et al. 2005). Similarly, institutional theory points out that the businesses adopt sustainability practices mainly to align their strategies with the requirements of institutions and not to meet strong sustainability (Bansal 2005). The natural resource-based theory shows that sustainability practices offer only competitive advantage for businesses (Hart 1995; Hahn et al. 2010). The main focus of these theories is on strategic management and suffer from the lack of estimating any trade-offs between three types of capital. Some of these methodologies have been based on general accepted guides (e.g., GRI and ISO 14031) to measure corporate triple-bottom-line performance (Azapagic 2003, 2004; Isaksson and Steimle 2009) while some methodologies have been focused on two aspects of sustainability (financial and environmental) to measure corporate sustainability (Burritt and Saka's 2006).

The other trend puts effort to integrate concepts of substitutability and rebound effect into composite indicators. The idea of sustainability could be an outstanding chance for businesses to align their strategies with the requirements of regulators which are close to the goals of strong sustainability (Epstein and Roy 2001). Some scholars suggest that a sustainable organization should face natural environmental as one of the significant stakeholders whose requests could be reasons which organization should address (Stubbs and Cocklin 2008).

Businesses community should follow a new scientific paradigm in order to change their behavior in the general ecological system and natural environment (Stead and Stead 2000). For this purpose, businesses community should adopt a new green management paradigm with goals like protection of carrying capacity of the planet. In the field of corporate environmental management, the concept of strong sustainability includes primarily the ability of businesses to preserve environment resources constant with previous years (Bebbington and Gray 1997).

Finally, capital theory is utilized to explain the basic procedures of businesses to the road of strong sustainability. Particularly, the constant substitution among three types of capital should be the core strategy of businesses (Dyllick and Hockerts 2002). Finally, governmental policies should support the business initiatives to sustainable development, while society will consume sustainable to motivate businesses' sustainable practices (Málovics et al. 2008).

3   Methodology

This chapter is based on a multi-step measurement approach (Weber 2008; Bai et al. 2012). First, it designed a conceptual model to describe the main steps of research structure. Second, each step is clearly analyzed (Nikolaou and Kazantzidis 2016). To this end, Fig. 1 depicts a research structure which consists of four main tasks. In the first task, the corporate strong sustainability definition has been made. In the second task, some details have been given in relation to three aspects of corporate sustainability. These definitions are useful for the next task in which indicators will be designed to measure economic, environmental, and social sustainability. The final task provides methodological details to integrate single sustainability indicators in a composite index to estimate a final score for strong sustainability performance of businesses.

Task 1: A Definition for Corporate Strong Sustainability (Authors' Own)
Van Marrewijk (2003) describes various definitions for corporate sustainability. Many definitions have been suggested for corporate sustainability with emphasis on different aspects of sustainability. Indeed, some of these definitions have been focused mainly on social aspects such as corporate citizenship and business ethics. They explain the responsible activities of businesses through their efforts to behave as good citizens by embracing the ethical values of society (Joyner and Payne 2002). A quite widespread socio-ethical model is the pyramid of Carroll which includes four basic responsibilities for businesses: economic, legislative, ethical, and philanthropic (Carroll 1991).

Other corporate sustainability definitions pay more attention to environmental issues. By following classical Brundtland's definition for sustainable development, Labuschagne and Brent (2005) define corporate sustainability as the strategy of businesses to meet the needs of current generation without compromising the ability of future generation to meet their needs. Finally, some definitions focus on economic aspect of corporate sustainability by examining the influence of sustainability strategy on profits and shareholder value (Godfrey et al. 2009). However, a lot of corporate sustainability definitions have recently focused on triple-bottom-line

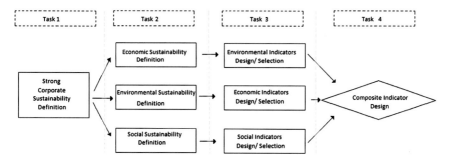

Fig. 1 The structure of the methodology

approach which implies simultaneously profits generation, progress in environmental performance, and ethical human resources management (Elkington 1998; Dyllick and Hockerts 2002).

Very little effort has been made on the field of corporate strong sustainability. The debate has limited more to ecological sustainability (van Weenen 1995; Wallner 1999) and failing to quantify thresholds for economic and social issues. The contemporary sustainable business models contribute to sustainable development in a relative way by measuring the continual improvement of sustainability performance among businesses (Málovics et al. 2008).

To this debate, a definition regarding strong sustainable business should be based on triple-bottom-line approach and well-defined thresholds for each aspect of sustainability. Therefore, sustainable businesses could be:

> these which invest successfully their profits and achieve simultaneously specific environmental and social objectives.

Task 2: Definitions of Three Aspects of Sustainability

In order this definition to be more useful, each aspect of sustainability should be measured. Hitherto, business economic sustainability is mainly measured through profits or shareholder values maximization. The profits cover various preconceptions since they are not able to respond clearly, if corporate sustainability practices improve profits or opposite. It is a high-risk approach to employ absolute figures of profits for measuring the financial performance of businesses because of its fluctuations might be a temporary consequence of the phase of business cycle. It is hard to describe an unsustainable period only using profits decreasing. Similarly, shareholder value is a limited indicator given that it covers only one group of stakeholders (shareholder) without examining many other necessary groups of stakeholders for sustainability performance. Thus, economic sustainability is defined as:

> the effective investment of financial capital of businesses in the long run period

The great part of current literature puts more emphasis on weak sustainability (free substitution among three types of capitals) and only a few studies examine environmental sustainability by suggesting renewable resources for firms. Bebbington and Gray (1997) have defined strong sustainability as the strategy of businesses to leave natural resources constant at an annual basis. Consequently, corporate strong environmental sustainability is defined as:

> their capability to work in an annually basis bellow specific thresholds which associated with basic environmental principles.

The last aspect, socially business sustainability focuses on protecting human rights and ethics. A classical definition for business social sustainability focuses on ethical issues of employees and needs of stakeholders (Wheeler et al. 2003; Aguilera et al. 2007). A distinction has been also made between the efforts of businesses to create social values both in inside and outside of business procedures. The social sustainability is defined as:

> the principles of equity and justice are successfully implemented by business community.

Task 3: Strong Sustainability Indicators Design

So far, there are two techniques to measure corporate sustainability performance. The former focuses on designing lists with single indicators suitable to measure each aspect of sustainability (Searcy 2012). The latter concentrates on developing composite indexes. Both techniques have some common flaws, especially, during the procedures of selecting indicators and connecting with the concept of strong sustainability (Singh et al. 2009; Hahn et al. 2010; Hediger 2010).

Many models have suggested determining the relationship between sustainability strategies and shareholders' value. The weaknesses of such financial indicators are focused first on the limited scope of shareholder idea (e.g., only a group of stakeholder idea) and second on unclear signal of profits as an absolute figure. Although many models have suggested determining the relationship between sustainability strategies and financial performance, nevertheless there are some weaknesses. One weakness is their focus only on the limited scope of shareholder idea (e.g., only a group of stakeholder idea). A second weakness is the unclear signal of profits as an absolute figure. A last but not least weakness of suggested financial indicators is the lack of consensus among scholars who provide various financial indicators (e.g., turnover ratio, average capital employed, total income or revenue, total costs, return on investment, and sales) and methodologies (Székely and Knirsch 2005; Singh et al. 2007; Dočekalová and Kocmanová 2016).

Taking into account these failures, it is suggested initially the Net Present Value (NPV) of profits (Table 1: Eq. 1) and furthermore, the classical Internal Rate of Return (IRR) to show the efficient way in which businesses invest their financial capital (Table 1: Eq. 2).

Additionally, in the context of strong sustainability, the financial aspect is necessary to be seen over a long-run period. To determine the rate of successful investment of firm, a threshold is necessary to be combined with IRR. Relative literature suggests Costs of Capital (CC) as a suitable indicator to compare with IRR. This combination shows the efficient investment of financial capital of

Table 1 Sustainability indicators under triple-bottom-line approach

Symbols	Equation	Number of equation	Details
NVP	$NPV = \frac{\sum_{i=1}^{t}(B_t - C_t)}{(1+r)^t} - II$	Eq. 1	Net present values B_t: benefits for year t, C_t: costs for year t, $B_t - C_t$: profits, r: discount rate, II: initial investment
IRR	$\frac{\sum_{i=1}^{t}(B_t - C_t)}{(1+IRR)^t} - II = 0$	Eq. 2	Internal rate of return
ECO_I$_t$	$ECO\_I_t = IRR_t - CC_t$	Eq. 3	Economic indicator in time t, IRR_t: internal rate of return, CC_t: costs of capital
ENV_I$_{i,t}$	$ENV\_I_{i,t} = Trh_{i,t} - f(EP_{i,t})$	Eq. 4	Environmental indicators i in time t, $Thr_{i,t}$: thresholds of i indicator in t time, $f(EP_{i,t})$: environmental performance of i indicator in t time
SOC_I$_{j,t}$	$SOC\_I_{j,t} = f(SP_{j,t}) - Trh_{j,t}$	Eq. 5	Social indicators of j indicator in t time, $Thr_{j,t}$: thresholds of j indicator in t time, $f(SP_{j,t})$: environmental performance of j indicator in t time

businesses. Eq. 3 (Table 1) shows Economic Indicator (ECO_I). This shows that when IRR is greater than CC, then ECO_I (Economic Indicator) contributes positive to composite sustainability indicator and negative in the opposite case.

In the second aspect of sustainability, many indicators have been suggested to measure the environmental dimension. A significant failure of such indicators is their limitation on measuring weak sustainability (Málovics et al. 2008). Although some efforts have been made to measure strong environmental sustainability, nevertheless, they have achieved small progress in the field (Figge and Hahn 2004; Velena et al. 2003).

To overcome some failures of the previous methodologies, it is suggested environmental indicators (ENvironmental Indicators—ENV_$I_{i,t}$) as combination of Environmental Performance (EP) of businesses for i indicator (e.g., air emission, water use, wastewater production) in t year with threshold for each indicator in t year (Table 1: Eq. 4). The influence of environmental indicators is zero on the composite index in the case where corporate environmental performance is equal to threshold. The business influence is positive in the composite index in the case where corporate environmental performance indicators are greater than thresholds and negative in the opposite case.

Similarly, Social Indicators (SOCial Indicators, SOC_$I_{i,t}$) is estimated as an abstraction between Thresholds (Thr$_{i,t}$) and Social Performance of firms (SP$_{i,t}$) (Table I: Eq. 5). These indicators have a positive influence on composite index in the case when business social performance is greater to threshold and negative in the opposite case.

Task 4: A Composite Index for Business Strong Sustainability

To integrate business sustainable indicators to a composite sustainability index, some further mathematical transformations are necessary such as scores normalization, weight factors estimation, and composite index equation design.

Equation 6 shows the idea of normalized indicators: It converts scores of business sustainability indicators in a scale between −1 and 1. The score of each aspect of sustainability is converted by utilizing a specific mathematical equation (Table 2: Eqs. 6.1–6.3). Second, suitable weight factors are determined which are necessary to integrate single indicators to the composite index which is summed up to the unit (Table 2: Eq. 8). Third, it proposed Eq. 7 (Table 2) to integrate normalized indicators into composite indicator.

Finally, social indicator indicates business contribution to strong sustainability and it is ranged from −1 to 1. Negative scores [−1,0) indicate the business contribution to weak sustainability, while scores greater or equal to zero [0,1] means that businesses contribute to strong sustainability.

4   A Case Study: An Evidence from Food Industry

A case study is analyzed to test the validity of the proposed methodology. It aims to indicate the practicability of methodology to strong sustainability of firms (Nikolaou et al. 2013). The sample includes three businesses in the food sector. For confidential reasons, it omitted the names of businesses and it utilized only symbols B_1 (for Business 1), B_2 (for Business 2), and B_3 (for Business 3.) Starting from the economic aspect of sustainability, Table 3 shows economic performance indicators for sampled businesses during a period of 5 years (data emerged from sustainability and annual reports of businesses). The score of indicators is based on Eqs. 1, 2, 3, and 6.1. The final column shows normalized economic indicators for each firm per year.

Table 4 analyzes the final environmental indicators in comparison to thresholds in order to show how businesses reach the goals of strong sustainability. The results show that the majority of environmental indicators are greater to thresholds. Only, 5% of indicators achieve scores under thresholds (Thr) and equal to zero are 7% of indicators.

Table 5 illustrates the final score of business environmental indicator which is estimated as a product between normalized environmental scores and weight factors. It is significant to say that weight factors are emerged from a questionnaire-based survey in five experts in the field of corporate sustainability. 32% of environmental indicators contribute to the overall sustainability score.

Similarly, social indicators have been quantified by information which is drawn from sustainability and annual reports. Table 6 shows social indicators as abstraction between social performance indicators and thresholds. The findings show that most of the social indicators are greater to thresholds. This means that

Table 2 Composite sustainability index

Symbols	Equation	Number of equation	Details
$I_{i,t:\text{norm}}$	$\begin{cases} \dfrac{x_{i,t}}{X_{\max,t}}, & X > 0 \\ 0, & X = 0 \\ -\dfrac{x_{i,t}}{X_{\max,t}}, & X < 0 \end{cases}$	Eq. 6	Normalized indicator i in period t X_i performance of firm $X_{\max,t}$ *is the max score of all indicators*
$ECO\_I_{i,t:\text{norm}}$	$\begin{cases} \dfrac{ECO\_I_{i,t}}{ECO\_I_{i,t,\max}}, & ECO\_I_i > 0 \\ 0, & ECO\_I_i = 0 \\ -\dfrac{ECO\_I_{i,t}}{ECO\_I_{i,t,\max}}, & ECO\_I_i < 0 \end{cases}$	Eq. 6.1	Normalized economic indicators $ECO\_I_{i,t}$ economic performance of business, $ECO\_I_{i,t,\max}$ the max score of economic performance
$ENV\_I_{i,t:\text{norm}}$	$\begin{cases} \dfrac{ENV\_I_{i,t}}{ENV\_I_{i,t,\max}}, & ENV\_I_{i,t} > 0 \\ 0, & ENV\_I_{i,t} = 0 \\ -\dfrac{ENV\_I_{i,t}}{ENV\_I_{i,t,\max}}, & ENV\_I_{i,t} < 0 \end{cases}$	Eq. 6.2	Normalized environmental indicators $ENV\_I_{i,t}$ environmental performance of business, $ENV\_I_{i,t,\max}$ the max score of environmental performance
$SOC\_I_{f,t:\text{norm}}$	$\begin{cases} \dfrac{SOC\_I_{f,t}}{SOC\_I_{f,t,\max}}, & SOC\_I_{f,t} > 0 \\ 0, & SOC\_I_{f,t} = 0 \\ -\dfrac{SOC\_I_{f,t}}{SOC\_I_{f,t,\max}}, & SOC\_I_{f,t} < 0 \end{cases}$	Eq. 6.3	Normalized social indicators $SOC\_I_{f,t}$ social performance of business, $ECO\_I_{i,t,\max}$ the max score of social performance
SBS_I_t	$\displaystyle\sum_{i=1}^{n} w_i ECO\_I_{i,t\text{normal}} + \sum_{l=n}^{m} w_l ENV\_I_{l,t\text{normal}} + \sum_{f=m}^{k} w_f SOC\_I_{f,t\text{normal}}$	Eq. 7	Strong business sustainability index Sum of economic, environmental and social indexes
WF	$\displaystyle\sum_{i=1}^{n} w_i + \sum_{l=n}^{m} w_l + \sum_{f=m}^{k} w_f = 1$	Eq. 8	Weight factors

Table 3 Economic indicators' values (authors' own)

Businesses	Years	B_t	C_t	$B_t - C_t$	r	$1 + r$	$(B_t - C_t)/(1 + r)$[b]	IRR[c]	ECO_I$_t$[d]	ECO_I$_{t,\text{normal}}$[e]
B_1[a]							−100,000			
	1	22,000	15,300	6700	0.6	0.4	16,750	0	0	0
	2	23,000	16,500	6500	0.6	0.16	40,625	−38%	−0.98	0.98
	3	25,000	18,000	7000	0.6	0.064	109,375	8%	−52%	0.43
	4	25,500	22,000	3500	0.6	0.0256	136,718.75	46%	−0.14	0.20
	5	26,000	23,500	2500	0.6	0.01024	244,140.6	68%	0.08	0
B_2[a]							−100,000			
	1	28,000	23,000	5000	0.6	0.4	12,500	0	0	0
	2	13,000	32,500	−19,500	0.6	0.16	−121,875	−8%	−0.52	−0.43
	3	35,000	32,000	3000	0.6	0.064	46,875	−1%	−0.59	0.36
	4	41,000	38,000	3000	0.6	0.0256	117,187.5	10%	−0.5	0.45
	5	41,500	37,500	4000	0.6	0.01024	390,625	10%	−0.5	0.45
B_3[a]							−100,000			
	1	30,000	23,000	7000	0.6	0.4	17,500	0	0	0
	2	31,000	24,500	6500	0.6	0.16	40,625	0	0	0
	3	32,500	28,800	3700	0.6	0.064	57,812.5	−8%	−0.52	0.43
	4	34,300	30,000	4300	0.6	0.0256	167,968.75	−8%	−0.52	0.43
	5	37,200	34,200	3000	0.6	0.01024	292,968.75	23%	−0.37	0.57

[a]Businesses: F_1, F_2, F_3
[b]Eq. 1
[c]Eq. 2
[d]Eq. 3
[e]Eq. 6.1

Table 4 Environmental indicator scores (authors' own)

ENV_I[b]	Trh[c]	Years														
		B_1^a					B_2^a					B_3^a				
		1	2	3	4	5	1	2	3	4	5	1	2	3	4	5
ENV_I1	0.54	0.12	0.03	0.05	0.03	-0.01	0.07	0.08	0.04	0.03	0.04	0.05	-0.01	0.05	0.06	0.05
ENV_I2	0.51	-0.12	0.02	0.01	0.03	0.03	0.04	0.03	0.02	0.01	0.01	0.02	0.04	0	0.02	0.03
ENV_I3	0.42	-021	0.01	0	0.02	0.03	0.01	0.02	0.02	0	0.02	0.02	0.01	0	0.04	0
ENV_I4	0.32	-0.31	0	0.01	-0.04	0	-0.01	0	0.02	-0.02	0	0	0.01	0	0.02	0
ENV_I5	0.41	0.22	0.01	0.03	0.04	0.02	0.03	0	0	0.01	0.02	0.03	0	0.03	0.01	0.02
ENV_I6	1220	200	100	0	100	-100	220	190	210	207	200	190	100	0	-100	100
ENV_I7	560	50	0	10	20	30	40	30	20	30	40	44	47	60	30	40
ENV_I8	3302	290	210	110	280	0	110	210	120	0	300	290	280	310	110	220
ENV_I9	36	5	3	4	2	4	3	2	1	6	4.5	3.5	2.5	7	5	4
ENV_I10	58	5	4	3	2	4	5	6	7	4.5	4	5	3	6	7	6
ENV_I11	27	5	3	2	1	0	3	2	4	3	2	1	3	6	4.5	4
ENV_I12	18	2	1	2	2	3	2	1	3	4	3	2.5	2	1	0.5	2
ENV_I13	14	1	0	-1	0	2	1	0	-0.5	-0.5	-1	-1.5	1	0	-1	-2
ENV_I14	16	1	0	0.5	0.7	0.8	1	0.8	0.7	0.5	0.4	0.3	1	0.7	0.8	0.9
ENV_I15	17	2	1.5	1.5	1.5	1	0	2	1.5	1.4	1.3	0	1	3	4	1
ENV_I16	19	1	0	2	0.5	0.9	2	1.5	1	0.5	0.2	0.1	0	-1	1	0
ENV_I17	212	12	0	20	9	7	5	0	-6	2	6	5	4	7	10	12
ENV_I18	0.48	0.21	0.21	0	0.22	0.23	0.22	0.21	0.22	0	0.23	0.22	0.24	-0.21	-0.22	0.22
ENV_I19	0.39	0.11	0.03	0.02	0.02	0.01	0.02	0.05	0.06	0.07	0.03	0.02	0.01	0	-0.01	0
ENV_I20	222	18	10	10	18	17	30	25	19	18	13	17	18	19	18	40
ENV_I21	62	12	8	12	13	14	9	8	11	11	9	8	9	10	9	11
ENV_I22	611	12	9	8	11	12	13	10	9	8	11	16	12	10	0	10

[a]Businesses 1, 2 and 3

[b]Environmental indicators

[c]Thresholds

Table 5 Final score of corporate environmental performance (authors' own)

	WF	B1[a]					B2[a]					B3[a]				
Businesses		1	2	3	4	5	1	2	3	4	5	1	2	3	4	5
ENV_I1[b]	0.02	-0.05	0.10	0.16	0.10	1.26	0.021	0.24	0.013	0.010	0.013	0.016	1.26	0.016	0.01	0.016
ENV_I2	0.01	0.1	0.09	0.06	0.02	0.012	0.015	0.02	0.009	0.006	0.006	0.009	0.05	0.01	0.09	0.03
ENV_I3	0.01	0.01	0.09	0.06	0.02	0.015	0.009	0.012	0.012	0.006	0.012	0.012	0.009	0.006	0.018	0
ENV_I4	0.02	0.01	0.01	0.07	7.36	0.014	-0.010	0.014	0.021	-0.07	0.014	-0.03	0.017	0.014	-0.021	-0.010
ENV_I5	0.09	0.05	0.03	0.07	0.09	0.005	0.007	0	0.001	0.003	0.005	0.007	0.001	0.007	0.003	0.005
ENV_I6	0.01	0.02	0.02	0	0.02	-0.012	0.026	0.023	0.025	0.025	0.024	0.023	0.012	0	-0.01	0.012
ENV_I7	0.01	0.02	0	0.02	0.05	0.007	0.010	0.007	0.005	0.007	0.010	0.011	0.011	0.015	0.007	0.010
ENV_I8	0.01	0.07	0.02	0.06	0.06	0	0.006	0.012	0.006	0	0.017	0.016	0.016	0.018	0.006	0.012
ENV_I9	0.02	0.04	0.07	0.01	0.03	0.010	0.007	0.003	0	0.017	0.012	0.008	0.005	0.021	0.014	0.010
ENV_I10	0.06	0.03	0.02	0.01	0	0.002	0.003	0.004	0.006	0.003	0.002	0.003	0.001	0.004	0.006	0.004
ENV_I11	0.09	0.07	0.04	0.03	0.01	0	0.004	0.003	0.006	0.004	0.003	0.001	0.004	0.009	0.006	0.006
ENV_I12	0.01	0.05	0.01	0.05	0.05	0.008	0.005	0.001	0.008	0.012	0.008	0.006	0.005	0.001	0	0.005
ENV_I13	0.01	0.01	0.009	0.04	0.09	0.018	0.013	0.009	-0.06	-0.006	-0.004	-0.002	0.013	0.009	-0.004	0
ENV_I14	0.02	0.02	0	0.01	0.04	0.016	0.021	0.016	0.014	0.010	0.008	0.006	0.021	0.014	0.016	0.019
ENV_I15	0.02	0.01	0.09	0.009	0.009	0.006	0	0.012	0.009	0.008	0.078	0	0.006	0.018	0.024	0.006
ENV_I16	0.02	0.01	0.09	0.02	0.03	0.017	0.027	0.022	0.018	0.013	0.010	0.01	0.009	0	0.018	0.009
ENV_I17	0.01	0.01	0.004	0.018	0.010	0.009	0.007	0.004	0	0.005	0.008	0.007	0.006	0.009	0.009	0.011
ENV_I18	0.02	0.01	0.01	0.07	0.04	0.017	0.014	0.010	0.014	0.007	0.176	0.014	0.021	-0.003	0	0.014
ENV_I19	0.02	0.06	0.02	0.09	0.09	0.006	0.009	0.018	0.021	0.024	0.012	0.009	0.006	0.003	0	0.003
ENV_I20	0.02	0.09	0	0	0.07	0.006	0.018	0.03	0.008	0.007	0.002	0.006	0.07	0.008	0.007	0.027
ENV_I21	0.01	0.04	0	0.08	0.10	0.012	0.002	0	0.006	0.006	0.002	0	0.02	0.004	0.002	0.006
ENV_I22	0.01	0.01	0.10	0.09	0.12	0.13	0.014	0.11	0.010	0.009	0.125	0.018	0.13	0.011	0	0.011

[a]Businesses 1, 2 and 3
[b]Environmental indicators

Table 6 Social indicators in according to the thresholds[a] (authors' own)

SOC_I_i[c]	Trh	B_1[b] Years					B_2					B_3				
		1	2	3	4	5	1	2	3	4	5	1	2	3	4	5
SOC_I_1	0.4	0.05	0.15	−0.5	0	−0.2	−0.1	0.05	0.1	0.2	0.07	0.06	0.04	−0.1	−0.08	−0.1
SOC_I_2	0.6	0.05	−0.1	−0.3	0.07	0.06	−0.2	0.06	−0.1	−0.9	−0.1	−0	−0.2	0.06	0.1	0.05
SOC_I_3	0.3	−0.01	0.01	0.04	0.01	0.3	−0.3	−0.1	−0.1	0.01	0	0.06	0.09	0.1	0.14	0.15
SOC_I_4	0.6	−0.01	−0.3	−0.2	0	0.01	0.02	0	−0.1	−0.1	−0.1	0.04	0.01	−0.1	−0.05	−0
SOC_I_5	0.4	−0.01	0.01	0.02	−0.01	0	−0.3	0.1	0.03	−0.5	0	0.01	0.02	0.04	0.05	0.06
SOC_I_6	7.1	−0.1	0.2	−0.3	−0.4	0	0.1	0.4	−0.4	−0.8	−0.7	−0.3	−0.2	−0.1	0	0.1
SOC_I_7	29	1	−1	−2	2	3	4	1	2	−1	−3	−1.5	0	1	2	1
SOC_I_8	0.2	−0.02	0.03	0.04	0.01	0	−0.1	0.03	−0	0.01	0.05	0.01	0	0.03	0.07	0.01
SOC_I_9	21	−1	1	2	3	−3	−2	−3	−0.5	−2	−1	0	1	−2	−1.5	0
SOC_I_{10}	0.4	−0.04	0.01	−0.2	−0.03	−0.1	0	−0	−0	−0.6	−0.1	−0.1	−0.8	−0	−0.02	−0
SOC_I_{11}	0.3	−0.01	0	0.02	0.01	0.06	0.05	0.02	0.01	0.02	0	27.7	−0.2	0	0.01	0.02
SOC_I_{12}	22	−2	3	5	2	1	0	1	0	−2.5	−3.6	1.2	2	3.1	6.2	7.3
SOC_I_{13}	43	−1	−1	0	1	2	4	−1	−2	0	−4	−5	−2	−2.5	−5	−6
SOC_I_{14}	11	−12	−11	−10	−10	−9	−11	−12	−7	−8	−10	−11	−11	−9	−7.5	−7
SOC_I_{15}	3.3	−0.3	0.2	0.1	0	0.3	0.4	0.5	0.2	0	0.1	0.3	0.1	0.4	0.6	0.3
SOC_I_{16}	34	−2	−1	1	2	3	4	3	2	1.6	−1.7	−1.6	−1.7	−1	0	1
SOC_I_{17}	4.3	−0.3	0.2	0	−0.1	0.2	0.1	0	−0.2	0.4	0.2	0.5	0.6	0.4	−0.1	0.1
SOC_I_{18}	3 .3	−0.1	−0.3	−0.2	0	0.2	0	0.3	0.1	0.4	0.4	0.5	0.1	0	0.1	−0.1
SOC_I_{19}	5.2	−0.2	−0.1	0	0.1	0	0.3	0.2	0.4	0.1	−0.1	0.5	0	0.3	−0.1	0.1
SOC_I_{20}	3.5	−0.5	−0.1	0	−0.2	−0.4	0.2	0.3	−0.1	−0.2	0.2	0.1	−0.2	−0.3	−0.4	−0.2
SOC_I_{21}	315	−150	−50	150	−120	50	−30	170	150	350	250	450	550	−50	50	150
SOC_I_{22}	1.5	−0.7	−0.2	0.3	−1.2	−0.4	−0.1	0	0.1	0.2	0	0.1	0.1	−0.3	−0.2	−0.5
SOC_I_{23}	0.3	−0.02	−0.1	−0.2	−0.02	−0.1	−0.1	−0	−0	−0.2	−0	−0	−.2	−0	−0.02	−0

(continued)

Table 6 (continued)

SOC_I_i^c	Trh^a	Businesses														
		Years														
		B_1^b					B_2					B_3				
		1	2	3	4	5	1	2	3	4	5	1	2	3	4	5
SOC_I_24	4.6	-0.6	-0.1	0	0.1	0.2	-0.3	-0.4	-0.1	-0.3	0.1	-0.1	-0.4	0	0.1	-0.4
SOC_I_25	1.3	-1.5	-1	-0.5	0.5	-2.5	-0.5	2.5	1.5	0.5	-1.5	-0.5	0	-0.5	3.5	0.5

[a]In according to Eq. 5
[b]Businesses 1, 2 and 3
[c]Social indicators which are described in appendix B

Table 7 Final score of social indicators[a]

SOC_I	WF	B_1^b					B_2					B_3				
		Years														
		1	2	3	4	5	1	2	3	4	5	1	2	3	4	5
SOC_I1	0.08	0.011	0.015	0.006	0.009	0	0.004	0.01	0.013	0.018	0.012	0.011	0.010	0.005	0.08	0.011
SOC_I2	0.01	0.016	0.010	0.008	0.018	0.017	0.009	0.017	0	0.002	0.004	0.008	0.009	0.017	0.021	0.016
SOC_I3	0.04	-0.06	-0.04	-0.01	-0.04	-0.02	-0.08	-0.04	-0.04	-0.07	-0.05	-0.09	-0.06	-0.05	0.04	-0.6
SOC_I4	0.07	0.019	0.016	0.018	0.021	0.022	0.024	0.021	0	0.003	0.007	0.027	0.022	0.009	0.07	0.019
SOC_I5	0.04	0.008	0.013	0.015	0.008	0.011	0.004	0.013	0.017	0	0.011	0.013	0.015	0.019	0.04	0.008
SOC_I6	0.01	0.012	0.017	0.008	0.007	0.014	0.015	0.021	0.007	0	0.001	0.008	0.010	0.012	0.01	0.012
SOC_I7	0.01	-0.002	0.021	0.033	-0.014	-0.026	-0.08	-0.02	-0.00	0.012	.045	0.027	0.009	-.02	0.01	-0.02
SOC_I8	0.08	0	0.010	0.012	0.006	0.004	0.002	0.010	0.002	0.006	0.014	0.006	0.004	0.010	0.08	0
SOC_I9	0.05	0.005	0.010	0.012	0.015	0	0.002	0	0.006	0.002	0.005	0.007	0.010	0.002	0.015	0.005
SOC_I10	0.01	0.002	0.004	0.003	0.002	0.003	0.004	0.003	0.003	0.001	0.001	0.005	0	0.002	0.01	0.002
SOC_I11	0.08	6.56	1.31	2.62	1.97	5.25	4.59	2.62	1.97	2.62	1.31	0.018	0	1.31	0.08	6.56
SOC_I12	0.01	0.003	0.012	0.016	0.010	0.008	0.007	0.008	0.007	0.002	0	0.009	0.010	0.013	0.01	0.003
SOC_I13	0.04	0.012	0.012	0.014	0.016	0.019	0.024	0.012	0.009	0.014	0.004	0.002	0.009	0.008	0.04	0.012
SOC_I14	0.01	0	0.004	0.008	0.008	0.012	0.004	0.002	0.021	0.016	0.008	0.006	0.004	0.012	0.01	0
SOC_I15	0.021	0.006	0.007	0.007	0.007	0.008	0.008	0.008	0.007	0.007	0.107	0.008	0.007	0.008	0.01	0.006
SOC_I16	0.08	0	0.003	0.009	0.012	0.015	0.018	0.015	0.012	0.010	0.000	0.001	0.009	0.003	0.08	0
SOC_I17	0.05	0	0.008	0.005	0.003	0.008	0.006	0.005	0.001	0.011	0.008	0.013	0.015	0.011	0.05	0
SOC_I18	0.08	0.004	0	0.002	0.006	0.011	0.006	0.013	0.009	0.015	0.015	0.018	0.009	0.006	0.08	0.004
SOC_I19	0.021	0	0.003	0.006	0.009	0.006	0.015	0.012	0.018	0.009	0.003	0.021	0.006	0.015	0.01	0
SOC_I20	0.04	0	0.012	0.015	0.009	0.003	0.021	0.024	0.012	0.009	0.021	0.018	0.009	0.006	0.04	0
SOC_I21	0.024	0	0.004	0.013	0.001	0.008	0.005	0.014	0.013	0.022	0.017	0.026	0.030	0.004	0.04	0

(continued)

Table 7 (continued)

SOC_I	WF	B₁[b]					B₂					B₃				
		1	2	3	4	5	1	2	3	4	5	1	2	3	4	5
SOC_I₂₂	0.027	0.009	0.018	0.027	0	0.014	0.02	0.021	0.023	0.025	0.021	0.023	0.023	0.016	0.07	0.009
SOC_I₂₃	0.07	0.015	0.017	0.011	0.007	0.023	0.019	0.027	0.023	0.011	0.007	0.016	0.011	0.015	0.07	0.015
SOC_I₂₄	0.024	0	0.015	0.018	0.021	0.024	0.009	0.006	0.015	0.009	0.021	0.015	0.006	0.018	0.04	0
SOC_I₂₅	0.01	0.003	0.005	0.007	0.016	0	0.007	0.01	0.014	0.010	0.003	0.007	0.008	0.007	0.01	0.003
Total		**0.11**	**0.23**	**0.27**	**0.18**	**0.20**	**0.19**	**0.27**	**0.21**	**0.21**	**0.23**	**0.31**	**0.249**	**0.23**	**0.58**	**0.276**

[a]In according to Eq. 5
[b]Businesses 1, 2 and 3

Table 8 Final composite businesses sustainability score

Year	SBS_I		
	B_1	B_2	B_3
1	0.11	0.43	0.50
2	0.36	0.40	0.43
3	0.45	0.42	0.41
4	0.41	0.40	0.74
5	0.39	0.45	0.47

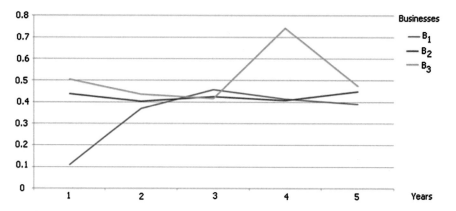

Fig. 2 Comparative analysis among SBS of the sampled businesses

social indicators have a significant contribution on the composite corporate sustainability index.

Table 7 shows the final scores of corporate social indicators. They are estimated as a product between weight factors and normalized social scores. The significance of these indicators is ranged from 20 to 30%.

Finally, Table 8 estimates composite sustainability score as the sum of three sub-indicators such as economic, environmental, and social indicator.

Finally, Fig. 2 shows a comparative analysis for three businesses. Particularly, it is showed the evolution for their composite business sustainability performance indexes. It is identified that the second business presents a constant increasing, while the first business achieves initially a short growth and afterwards a sharply decrease of total sustainability performance.

5   Final Remarks

This book chapter suggests a methodology to measure business contribution to strong sustainability. It focuses on developing a composite business sustainability index. The first innovation of the proposed methodology is the new indicators which combine the performance of businesses with specific thresholds. This could offer a clear signal to stakeholders regarding the contribution of businesses to strong sustainability. The second innovation is the clarification of each aspect of sustainability in an operational manner.

Starting from the financial aspect, it has been made an effort to integrate some of the considerable knowledge of sustainability into financial indicators such as long-run viability and equity among generations. A significant contribution to the debate of financial sustainability is the idea of efficiency behind the financial indicators which assist in overcoming recent analysis for economic sustainability only as profit maximization. This seems to be unable to explain temporary losses in financial statements. Furthermore, the worry for persistent progress of profits cannot promise continual business sustainability due to the fact that sustainability is a complex problem with many different variables. Continues growth for profits is a permanent request of shareholders who endeavors to increase their earnings per share in a daily basis.

The proposed framework contributes also by developing specific thresholds for each aspect of sustainability. They assist in estimating the proper level of participation of businesses to the overall strong sustainability performance. This assists in overcoming the request of current measurement techniques which propose only the continual improvement of each aspect without taking into account the certain goals of sustainability such as safe minimum standards and protection of carrying capacity of ecosystems. The contribution to this debate is made by suggesting environmental thresholds which are necessary to be associated with views such as carrying capacity, safe minimum standards, and critical capital.

Finally, the case study provides practical implications of the proposed methodology for scholars and stakeholders. The diagrammatically representations of corporate sustainability performance provide an easy instrument to make comparative analysis of business sustainability. The suggested methodological framework contributes to food industry literature by integrating the concept of strong sustainability into the overall debate of food industry sustainability. So far, two interesting academic debates are in this field such as Corporate Social Responsibility of food industry and sustainability of supply chain of food industry. The former debate emphasizes on examining the responsibilities of food industry against the health of consumers by analyzing many practices which are adopted or should be adopted by food industry (e.g., production of healthy and low-fat products as well as awareness of consumers regarding products' potential negative impacts). Food industry adopts CSR and sustainability practices to address the community's criticism regarding food production, distribution, and pricing. The latter debate focuses on examining sustainability in food supply chain and mainly the impacts of product distribution to sustainable development (e.g., energy and water consumption). A higher emphasis has been put on waste management of food industry in order to protect natural resources and guarantee food for future societies.

However, the suggested framework comes to contributes to these debates by doing more operational and quantifiable the concept of sustainability and mainly in the field of food manufacturing. It also contributes with thresholds which are associated with carrying capacity and capabilities of an area. The majority of previous methodologies offer various indicators under an unsystematic manner which confuse stakeholders to understand their content and make a similar indicator. The proposed framework is based on GRI guidelines to overcome this significant problem.

Certainly, as any other study, it has some limitations which could be a good basis for future research. The first limitation is associated with the size of the sample. Only three businesses is a small sample which should be increased in the future. Many other studies could be conducted regarding food industry. The second limitation pertains to the calculation of thresholds which should be made more clearly and accurately by estimating all supply chains of food industry. It could be a good area for future research by combining many different academic fields.

References

Aguilera, R. V., Rupp, D. E., Williams, C. A., & Ganapathi, J. (2007). Putting the S back in corporate social responsibility: A multilevel theory of social change in organizations. *Academy of Management Review, 32*(3), 836–863.

Aras, G., & Crowther, D. (2008). Governance and sustainability: An investigation into the relationship between corporate governance and corporate sustainability. *Management Decision, 46*(3), 433–448.

Atkinson, G. (2000). Measuring corporate sustainability. *Journal of Environmental Planning and Management, 43*(2), 235–252.

Azapagic, A. (2003). Systems approach to corporate sustainability: A general management framework. *Process Safety and Environmental Protection, 81*(5), 303–316.

Azapagic, A. (2004). Developing a framework for sustainable development indicators for the mining and minerals industry. *Journal of Cleaner Production, 12*(6), 639–662.

Bai, C., Sarkis, J., Wei, X., & Koh, L. (2012). Evaluating ecological sustainable performance measures for supply chain management. *Supply Chain Management: An International Journal, 17*(1), 78–92.

Bansal, P. (2005). Evolving sustainably: A longitudinal study of corporate sustainable development. *Strategic Management Journal, 26*(3), 197–218.

Baumgartner, R. J., & Ebner, D. (2010). Corporate sustainability strategies: sustainability profiles and maturity levels. *Sustainable Development, 18*(2), 76–89.

Bebbington, J., & Gray, R. (1997). Sustainable development and accounting: incentives and disincentives for the adoption of sustainability by transnational corporations. In C. Hibbitt & H. Blokdjik (Eds.), *Environmental accounting and sustainable development: The final report.* Amsterdam: Limperg Instituut.

Brattebø, H. (2005). Toward a methods framework for eco-efficiency analysis? *Journal of Industrial Ecology, 9*(4), 9–11.

Burritt, R. L., & Saka, C. (2006). Environmental management accounting applications and eco-efficiency: Case studies from Japan. *Journal of Cleaner Production, 14*(14), 1262–1275.

Carroll, A. B. (1991). The pyramid of corporate social responsibility: Toward the moral management of organizational stakeholders. *Business Horizons, 34*(4), 39–48.

Delai, I., & Takahashi, S. (2011). Sustainability measurement system: A reference model proposal. *Social Responsibility Journal, 7*(3), 438–471.

Dočekalová, M. P., & Kocmanová, A. (2016). Composite indicator for measuring corporate sustainability. *Ecological Indicators, 61,* 612–623.

Dyllick, T., & Hockerts, K. (2002). Beyond business case for corporate sustainability. *Business Strategy and the Environment, 11,* 130–141.

Elkington, J. (1998). Partnerships from cannibals with forks: The triple bottom line of 21st-century business. *Environmental Quality Management, 8*(1), 37–51.

Epstein, M. J., & Roy, M. J. (2001). Sustainability in action: Identifying and measuring the key performance drivers. *Long Range Planning, 34*(5), 585–604.

Faucheux, S., & O'Connor, M. (1998). Weak and strong sustainability. In S. Faucheux & M. O'Connor (Eds.), *Valuation for sustainable development: Methods and policy indicators.* Cheltenham: Edward Elgar.

Figge, F., & Hahn, T. (2004). Sustainable value added—Measuring corporate contributions to sustainability beyond eco-efficiency. *Ecological Economics, 48*(2), 173–187.

Figge, F., & Hahn, T. (2005). The cost of sustainability capital and the creation of sustainable value by companies. *Journal of Industrial Ecology, 9*(4), 47–58.

Figge, F., Hahn, T., Schaltegger, S., & Wagner, M. (2002). The sustainability balanced scorecard–linking sustainability management to business strategy. *Business Strategy and the Environment, 11*(5), 269–284.

Godfrey, P. C., Merrill, C. B., & Hansen, J. M. (2009). The relationship between corporate social responsibility and shareholder value: An empirical test of the risk management hypothesis. *Strategic Management Journal, 30*(4), 425–445.

Goyal, P., Rahman, Z., & Kazmi, A. A. (2013). Corporate sustainability performance and firm performance research: Literature review and future research agenda. *Management Decision, 51* (2), 361–379.

Hahn, T., Figge, F., Pinkse, J., & Preuss, L. (2010). Trade-offs in corporate sustainability: You can't have your cake and eat it. *Business Strategy and the Environment, 19*(4), 217–229.

Hart, S. L. (1995). A natural-resource-based view of the firm. *Academy of Management Review, 20* (4), 986–1014.

Hediger, W. (2010). Welfare and capital-theoretic foundations of corporate social responsibility and corporate sustainability. *The Journal of Socio-Economics, 39*(4), 518–526.

Hubbard, G. (2009). Measuring organizational performance: beyond the triple bottom line. *Business Strategy and the Environment, 18*(3), 177–191.

Hutchins, M. J., & Sutherland, J. W. (2008). An exploration of measures of social sustainability and their application to supply chain decisions. *Journal of Cleaner Production, 16*(15), 1688–1698.

Ilinitch, A., Soderstrom, N., & Thomas, T. (1998). Measuring corporate environmental performance. *Journal of Accounting and Public Policy, 17,* 383–408.

Isaksson, R., & Steimle, U. (2009). What does GRI-reporting tell us about corporate sustainability? *The TQM Journal, 21*(2), 168–181.

Joyner, B. E., & Payne, D. (2002). Evolution and implementation: A study of values, business ethics and corporate social responsibility. *Journal of Business Ethics, 41*(4), 297–311.

Labuschagne, C., & Brent, A. C. (2005). Sustainable Project Life Cycle Management: the need to integrate life cycles in the manufacturing sector. *International Journal of Project Management, 23,* 159–168.

Lozano, R., Carpenter, A., & Huisingh, D. (2015). A review of 'theories of the firm' and their contributions to corporate sustainability. *Journal of Cleaner Production, 106,* 430–442.

Málovics, G., Csigéné, N. N., & Kraus, S. (2008). The role of corporate social responsibility in strong sustainability. *The Journal of Socio-Economics, 37*(3), 907–918.

Montiel, I. (2008). Corporate social responsibility and corporate sustainability: Separate pasts, common futures. *Organization & Environment, 21*(3), 245–269.

Nikolaou I. E. (2017). A framework to explicate the relationship between CSER and financial performance: An intellectual capital-based approach and knowledge-based view of firm. *Journal of the Knowledge Economy,* in press.

Nikolaou, I., & Evangelinos, K. (2012). Financial and non-financial environmental information: significant factors for corporate environmental performance measuring. *International Journal of Managerial and Financial Accounting, 4*(1), 61–77.

Nikolaou, I. E., Evangelinos, K. I., & Allan, S. (2013). A reverse logistics social responsibility evaluation framework based on the triple bottom line approach. *Journal of Cleaner Production, 56,* 173–184.

Nikolaou, I. E., & Kazantzidis, L. (2016). A sustainable consumption index/label to reduce information asymmetry among consumers and producers. *Sustainable Production and Consumption, 6,* 51–61.

Nikolaou, I. E., & Matrakoukas, S. I. (2016). A framework to measure eco-efficiency performance of firms through EMAS reports. *Sustainable Production and Consumption, 8,* 32–44.

Olsthoorn, X., Tyteca, D., Wehrmeyer, W., & Wagner, M. (2001). Environmental indicators for business: A review of the literature and standardisation methods. *Journal of Cleaner Production, 9*(5), 453–463.

Rahdari, A. H., & Rostamy, A. A. A. (2015). Designing a general set of sustainability indicators at the corporate level. *Journal of Cleaner Production, 108,* 757–771.

Salvati, L., & Zitti, C. (2009). Substitutability and weighting of ecological and economic indicators: Exploring the importance of various components of a syn-thetic index. *Ecological Economics, 68,* 1093–1099.

Salzmann, O., Ionescu-Somers, A., & Steger, U. (2005). The business case for corporate sustainability: literature review and research options. *European Management Journal, 23*(1), 27–36.

Schaltegger, S., Lüdeke-Freund, F., & Hansen, E. G. (2012). Business cases for sustainability: the role of business model innovation for corporate sustainability. *International Journal of Innovation and Sustainable Development, 6*(2), 95–119.

Schmidt, I., Meurer, M., Saling, P., Kicherer, A., Reuter, W., & Gensch, C. O. (2004). Managing sustainability of products and processes with the socio-eco-efficiency analysis by BASF. *Greener Management International, 45,* 79–94.

Searcy, C. (2009). Setting a course in corporate sustainability performance measurement. *Measuring Business Excellence, 13*(3), 49–57.

Searcy, C. (2011). Updating corporate sustainability performance measurement systems. *Measuring Business Excellence, 15*(2), 44–56.

Searcy, C. (2012). Corporate sustainability performance measurement systems: A review and research agenda. *Journal of Business Ethics, 107*(3), 239–253.

Shwartz, M., Burgess, J. F., & Berlowitz, D. (2009). Benefit-of-the-doubt approaches for calculating a composite measure of quality. *Health Services and Outcomes Research Methodology, 9,* 234–251.

Singh, R. K., Murty, H. R., Gupta, S. K., & Dikshit, A. K. (2007). Development of composite sustainability performance index for steel industry. *Ecological Indicators, 7*(3), 565–588.

Singh, R. K., Murty, H. R., Gupta, S. K., & Dikshit, A. K. (2009). An overview of sustainability assessment methodologies. *Ecological Indicators, 9*(2), 189–212.

Sridhar, K., & Jones, G. (2013). The three fundamental criticisms of the triple bottom line approach: An empirical study to link sustainability reports in companies based in the Asia-Pacific region and TBL shortcomings. *Asian Journal of Business Ethics, 2*(1), 91–111.

Stead, J. G., & Stead, E. (2000). Eco-enterprise strategy: standing for sustainability. *Journal of Business Ethics, 24*(4), 313–329.

Steurer, R., Langer, M. E., Konrad, A., & Martinuzzi, A. (2005). Corporations, stakeholders and sustainable development I: A theoretical exploration of business–society relations. *Journal of Business Ethics, 61*(3), 263–281.

Stubbs, W., & Cocklin, C. (2008). Conceptualizing a "sustainability business model". *Organization & Environment, 21*(2), 103–127.

Székely, F., & Knirsch, M. (2005). Responsible leadership and corporate social responsibility: Metrics for sustainable performance. *European Management Journal, 23*(6), 628–647.

Tyteca, D., Carlens, J., Berkhout, F., Hertin, J., Wehrmeyer, W., & Wagner, M. (2002). Corporate environmental performance evaluation: Evidence from the MEPI project. *Business Strategy and the Environment, 11*(1), 1–13.

Van Marrewijk, M. (2003). Concepts and definitions of CSR and corporate sustainability: Between agency and communion. *Journal of Business Ethics, 44*(2), 95–105.

Van Passel, S., Nevens, F., Mathijs, E., & Van Huylenbroeck, G. (2007). Measuring farm sustainability and explaining differences in sustainable efficiency. *Ecological Economics, 62* (1), 149–161.

van Weenen, J. C. (1995). Towards sustainable product development. *Journal of Cleaner Production, 3,* 95–100.

Veleva, V., Hart, M., Greiner, T., & Crumbley, C. (2003). Indicators for measuring environmental sustainability: A case study of the pharmaceutical industry. *Benchmarking: An International Journal, 10*(2), 107–119.

Wallner, H. P. (1999). Towards sustainable development of industry: Networking, complexity and eco-clusters. *Journal of Cleaner Production, 7,* 49–58.

Wartick, S. L., & Cochran, P. L. (1985). The evolution of the corporate social performance model. *Academy of Management Review, 10,* 758–769.

Weber, M. (2008). The business case for corporate social responsibility: A company-level measurement approach for CSR. *European Management Journal, 26,* 247–261.

Wheeler, D., Colbert, B., & Freeman, R. E. (2003). Focusing on value: Reconciling corporate social responsibility, sustainability and a stakeholder approach in a network world. *Journal of General Management, 28*(3), 1–28.

Bio-compounds Production from Agri-food Wastes Under a Biorefinery Approach: Exploring Environmental and Social Sustainability

Sara González-García, Patricia Gullón and Beatriz Gullón

Abstract The society and industrial sectors are facing important challenges regarding the production of bioproducts influenced by social responsibility and environmental consequences. Biorefinery development reports two important goals in the transition towards a bio-based economy: (i) the displacement of fossil-based products by biomass-based ones and (ii) the setting up of a strong bio-based industry. In this sense, research is being addressed into bio-based products opportunities from biomass residues with the aim of obtaining promising building blocks and high-added value products. Environmental and economic analysis of some bioproducts can be found in the literature. However, the social dimension of sustainability is regularly forgotten although many attempts have been performed to standardise and provide the procedures to assess the social dimension. This chapter presents the production of potential bioproducts from agri-food industrial sector and assesses their sustainability from environmental and social perspectives with the aim of identifying potential hotspots. Since the methodology to assess environmental consequences is well known and standardised, special attention is paid on the selection of the social indicators considered for analysis. To do so, social impact assessment is conducted through involved stakeholders, surveys and field experiments. Thus, the methodology to assess the social dimension has been formulated in detail considering very different social well-being-based indicators.

Keywords Environmental profile · Life-cycle assessment · Oligosaccharides
Social impacts · Succinic acid

S. González-García (✉) · B. Gullón
Department of Chemical Engineering, School of Engineering,
Universidade de Santiago de Compostela, 15782 Santiago de Compostela, Spain
e-mail: sara.gonzalez@usc.es

P. Gullón
Chemical and Environmental Engineering Department,
University of the Basque Country, 20018 San Sebastián, Spain

© Springer Nature Singapore Pte Ltd. 2019
S. S. Muthu (ed.), *Quantification of Sustainability Indicators in the Food Sector*,
Environmental Footprints and Eco-design of Products and Processes,
https://doi.org/10.1007/978-981-13-2408-6_2

1   Introduction

The public environmental awareness, the outstanding increase in population and the concerns with negative environmental impacts from conventional products derived from fossil sources have prompted a necessity for more sustainable production systems. Sustainable development is considered one of the most challenging policy concepts ever developed (Spangenberg 2004). Despite the sustainability concept is open to debate (Bond and Morrison-Saunders 2011), three dimensions should be addressed that are environmental protection, economic growth and social equality (Iribarren et al. 2016) in agreement with the United Nations (2015). However, studies consider that the time perspective should be considered. It should be in line with Griggs and colleagues (Griggs et al. 2013) who suggested reformulating the definition of sustainable development reported by the Brundtland Commission as '*development that meets the needs of the present while safeguarding Earth's life-support system, on which the welfare of current and future generations depends*'. Therefore, environmental conditions have to be identified that facilitate successful human development and establish acceptable ranges for the biosphere to remain in that state.

The integration of environmental and economic analysis is known as eco-economy or eco-efficiency and it is related with the use of energy and resources in an efficient way. Iribarren and colleagues establish that although eco-efficiency is associated with sustainability (ISO 2012), other issues are also connected such as consumption patterns and population development (Iribarren et al. 2016).

However, the social dimension of sustainability is commonly forgotten although many efforts have been performed to standardise and provide the procedures to include it under the multidimensionality of life-cycle assessment (Norris et al. 2012; Norris 2014). Socio-economy and Corporate Social Responsibility (CSR) of an institution are closely linked involving aspects such as the establishment of security policies, social and community aids and programs of employee training, among others (Searcy 2012; Blaga 2013), which could be related to the achievement of a socio-environmental equilibrium. Nevertheless, social and environmental dimensions are considered contrasting aims, since social sustainability requires a minimum of economic growth whereas environmental sustainability sets an upper limit to this growth. Moreover, economic growth increases the average income but it does not automatically reduce the inequalities in the society (Spangenberg 2004). For this reason, the adopted policies should focus on the enhancement of existing synergies and the development of balanced criteria to avoid overemphasizing one dimension with respect to the other (Spangenberg 2004).

Although many methods are available in the literature to assess the sustainability (Jørgensen et al. 2008; Norris et al. 2012, 2014), they cannot be applied in a generic way. Moreover, different approaches can be identified according to the impact categories and/or indicators (midpoint or endpoint) included (Jørgensen et al. 2008). Therefore, research is being carried out in order to go beyond the traditional three pillars and define the sustainability in terms of alternative or additional dimensions. This is the case of the Prosuite project (Blok et al. 2013) set up by the

European Commission as part of the Seventh Framework program, in which five dimensions have been proposed to assess the sustainability of new technologies: human health, social well-being, prosperity, natural environment and exhaustible resources from a Life-Cycle Assessment (LCA) approach. According to it, not only quantitative but also qualitative indicators for environmental and social analysis should be managed.

Among the societal challenges of the Horizon 2020, the bio-based industries are essential elements of the European economic, environmental and societal policy. These challenges involve the transition from fossil-based European industries towards bioresource-based ones. For this, it is necessary that the substitution of conventional industrial processes and products by environmentally friendly bio-based ones, the development of integrated biorefineries and the search of potential markets for bio-based products. These industries will contribute to the sustainability development through the production systems with enhanced ecosystem services, zero waste and adequate societal value contributing.

The project Biorefinery Euroview has concretised the biorefinery concept as follows: '*Biorefineries could be described as integrated biobased industries using a variety of technologies to make products such as chemicals, biofuels, food and feed ingredients, biomaterials, fibres and heat and power, aiming at maximizing the added value along the three pillars of sustainability (Environment, Economy and Society)*'.

The development of an integrated biorefinery based on biowastes would allow to accomplish some of the five dimensions considered on the sustainable development: (i) environmental sustainability, through the valorisation whole residues based on the zero-waste concept; (ii) economic sustainability, by means of manufacturing multiple high-added value products which could enter different markets; (iii) social sustainability, since this approach would benefit different sectors by means of the creation of qualified jobs and covering the increasing demands of consumers in bio-based products (Fava et al. 2015).

The agri-food processing industries generate huge amounts of wastes which constitute a cheap source of high-added value bio-compounds, bio-based chemicals and biofuels.

Increasing the use of agri-food wastes can derive into a range of challenges such as the no competition for land for food production or urban expansion, the reduced impact on ecosystems and biodiversity, the no-contribution to water scarcity and no impact in food prices.

In this sense, the industrial sector is putting its attention into organic waste valorisation boosted by the necessity to transform the fossil-based production processes towards the bioresource-based ones. There is a series of chemicals which can be obtained from the agri-food wastes following the biorefinery concept and that are considered top by the US Department of Energy (2004). Among these top chemicals, the succinic acid ($C_4H_6O_4$) was identified as one of the ten 'top' chemicals that could be obtained from biorefinery carbohydrates (Bozell and Petersen 2010; Werpy and Petersen 2004). Succinic acid or butanedioic acid is a striking renewable platform chemical mostly due to its functionality and valuable derivatives (López-Garzón et al. 2014).

A biorefinery that supplements its production processes of low-value biofuels with high-value bio-based chemicals can facilitate efforts to moderate non-renewable fuel consumption while simultaneously providing the necessary financial motivation to encourage growth of the biorefining industry. However, social dimension is not always considered. Therefore, the inclusion of social aspects into the environmental life-cycle assessment (LCA) of products and systems is mandatory (Jørgensen et al. 2008), ensuring safe working conditions and respecting for workers' rights. Figure 1 schematically displays some interrelations in the value chain of sustainably integrated biorefinery systems. According to it, the social dimension in biorefinery systems creates jobs, improve health and provide leisure for the society (Budzianowski and Postawa 2016).

The study aims to deliver a broad LCA framework, taking into account the two pillars of sustainability, i.e. environmental and social. Thus, an integrated methodology from a holistic approach that supports decisions that product developers, policymakers and businesses must perform is proposed for the sustainability analysis. To do so, the methodology has been applied to specific case studies of bio-based products with current social and industrial interest: oligosaccharides and antioxidants with an important interest in food and pharmaceutical sectors as well as succinic acid, one of the platform chemicals labelled as top chemical.

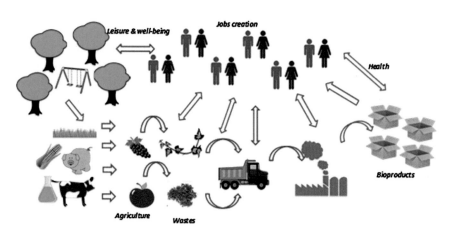

Fig. 1 Social and environmental interrelations in the assessment of biorefinery systems. Adapted from Budzianowski and Postawa (2016)

2　Top Bioproducts with Social Interest

Biomass resources can be considered as potential raw materials to produce high-added value products. In this sense, special attention is being paid in promising sugar-based chemicals and materials which could be considered as an economic driver for a biorefinery concept.

In the recent years, numerous studies and research activities are being carried out regarding the production of novel products with promising market potential and considered substitutes for existing petrochemicals as well as platform chemicals. This is the case of bio-based chemicals derived from carbohydrates (Bozell and Petersen 2010) and saccharides extracted from the hemicellulosic fraction (Gullón et al. 2018). Thus, the attention in this chapter has been focused on both bio-products categories which are also supported by the society's interest on their production.

Nowadays, it is not only important for the companies to improve the efficiency in their manufacturing processes with the aim of reducing waste production and resources consumption but also to increase the safety of workers, customers and environment as well as to offer in the markets novel products with the same or even improved properties than conventional ones.

2.1　Food Industry

The agricultural food processing generates more than 250 mil MT/year of by-products including stems, leaves, seeds, shells, pomace, bran, besides of food that do not meet the quality standards and do not make it into the production chain (Panouillé et al. 2007; Fava et al. 2015). In the past, these by-products were directly disposal to land, or were used for low added-value applications (composting or animal feed), causing environmental problems and also negative impacts on the sustainability of the food sector (Sala et al. 2017). Nowadays, the obtaining of different high-value products from agri-food by-products is an important ongoing field of research. These by-products are excellent sources of bioactive compounds including dietary fibre, antioxidants, oligosaccharides, vitamins, pectin, enzymes, pigments, organic acids *inter alia*, of special interest for the food industry (Galanakis 2012; Banerjee et al. 2017; Sagar et al. 2017).

These phytochemicals show interesting properties such as prebiotic, antibacterial, antihypertensive, antioxidant and cardioprotective capacity, and its consumption is related with beneficial effects to health or reduction the risk of diseases (Gullón et al. 2014; Muthaiyan et al. 2012). In the past few years, consumers have a higher health interest and more concern for the life quality, boosting the market of these natural biomolecules.

It is important to highlight that the use of agri-food by-products for obtaining different bioactive compounds is key to improve the environmental and economical sustainability of food sector (Fava et al. 2015; Alañón et al. 2017).

In this section, the obtaining of several bioactive compounds from agri-food by-products is described.

An interesting group of compounds that have seen increased its demand by the consumers in the last decade is integrated by pectin, oligosaccharides and dietary fibre which are recognised by their beneficial effects already listed above. Apple and citrus processing industries produce big amounts of wastes with a great potential to be valorised since they contain a wide variety of these bioactive compounds (Ndayishimiye and Chun 2017). Several research works have focused on obtaining these bio-compounds from apple pomace and citrus peels (Gullón et al. 2011; Gómez et al. 2013; Wang et al. 2014; Li et al. 2014). Apart from citrus peels and apple pulp, other agro-industrial sub-products such as sugar beet pulp, peach peels or pulps of grapes and pumpkin have also been found to contain high amounts of pectin and dietary fiber (Martínez et al. 2009; Müller-Maatsch et al. 2016; Banerjee et al. 2017; Sagar et al. 2017). Another interesting by-product for obtaining pectin is the one generated in the exotic fruits processing industry (Ayala-Zavala et al. 2011). Different industrial wastes associated with the production of olive oil were also evaluated for production of different mixtures of oligosaccharides with prebiotic potential (Fernández-Bolaños et al. 2004; Ruiz et al. 2017). With regard to the processing of cereals, it is worth highlighting the by-products generated in the grinding of wheat and de-hulling of rice, which are rich in dietary fibres including glucuronoarabinoxylans (Gullón et al. 2010, 2014; Hollmann and Lindhauer 2005).

The phenolic compounds are another important group of bioactive molecules which are present in large quantities in a great variety of agri-food by-products (Mohdaly et al. 2010; Ribeiro da Silva et al. 2014; Ndayishimiye and Chun 2017). Polyphenols have received a great deal of attention because of their capacity to combat the generation of free radicals in vivo, preventing cell damage and oxidative stress. Antioxidants also play a role as preservatives due to their ability to scavenge free radicals and prevent oxidation reactions in food (Deng et al. 2012; Banerjee et al. 2017). In this sense, over the last few decades, there has been a growing interest towards the production of natural antioxidants as an alternative to the less-safe synthetic antioxidants. Thus, it has been promoted that the obtaining of these phytochemicals from residues are derived from the food sector (Moreira et al. 2016). In fact, the peels of several fruits (e.g. apple, citrus, banana, watermelon, peaches and pineapple) are an excellent source of natural antioxidants such as phenolic acids, flavonoids, flavonols, catechins, tannins, procyanidins, anthocyanins, among others (Marín et al. 2007; Sagar et al. 2017; Banerjee et al. 2017). Grape skins and seeds, by-products of the juice and wine industries, are also rich sources of antioxidants compounds (González-Paramás et al. 2004; Teixeira et al. 2014; Barba et al. 2016). The tropical exotic fruit by-products contain also a great variety of antioxidant compounds that could be used for the formulation of nutraceuticals (Ayala-Zavala et al. 2011). Other successful example of by-product that can show a good profitability for the extraction of phenolic compounds is olive-derived biomass (Nadour et al. 2012; Lama-Muñoz et al. 2014; Ruiz et al. 2017). The potato processing industry also generates significant amounts of waste that have been investigated for the extraction of phenolics compounds (Singh et al. 2011; Sabeena Farvin et al. 2012).

In addition, the recovery of different pigments that can be used as food coloring agents have been widely investigated using fruits processing wastes (Ayala-Zavala et al.2011). Anthocyanins can be efficiently extracted from grape pomace or banana bracts (Monrad et al. 2014; Pazmiño-Durán et al. 2001). Citrus peel has also been utilised for the recovery of carotenoids (Ndayishimiye and Chun 2017; Agócs et al. 2007). Tomato industries produce large amounts of a by-product, known as tomato pomace, consisting mainly of skins and seeds (Lenucci et al. 2013). This by-product is an excellent source for the extraction of lycopene (Seifi et al. 2013; Baysal et al. 2000), and in recent years, this commercial pigment has received significant attention because of their important health benefits (Seifi et al. 2013; Lavelli and Torresani 2011).

Another interesting alternative for the valorisation of the agri-food by-products is the production of different enzymes through fermentation processes (Padma et al. 2012; Sandhya and Kurup 2013). Several research works have been focused in the use of many agro-industrial wastes for the production of various important enzymes in food industries. The apple pulp has a high content in cellulose and pectin, so it is an adequate substrate for the production of cellulase and polygalacturonase (Vilas-Boas et al. 2002; Zheng and Shetty 2000). Others pectin-rich fruit by-products like banana peel, orange peel, mango peel and pineapple peel have also been evaluated for polygalacturonase production (Padma et al. 2012). Potato peel has been also assessed to produce cellulases and amylases (dos Santos et al. 2012; Mushtaq et al. 2017).

The fruit by-products also can serve for the extraction of flavours and aromas (Sagar et al. 2017). One of these aromas is the vanillin which is widely used in the food industry. The microbial bio-transformation of pineapple wastes has been reported for the production of vanillin (Lun et al. 2014). Essential oils can be also obtained from of citrus peel or grape seeds.

2.2   Chemical Industry

Building blocks, reagents, intermediates … are classifications of chemicals with interest for industry and society, which could be obtained under a green approach. Petroleum, natural gas, sulphur dioxide, nitrogen and oxygen are considered raw materials for the production of commodity chemicals and intermediates destined to final products such as preservatives, fertilisers, dyes, food packaging and pharmaceuticals. However, all of them could be obtained from biomass feedstocks (starch, cellulose, hemicellulose, lignin, oil and protein) under a value chain approach (Werpy and Petersen 2004).

The use of carbohydrates as starting materials for chemicals production is well supported (Kam 2009). The lignocellulosic feedstock-based biorefinery, also known as green biorefinery, is being favoured in research, development and industrial implementation since the production of lignocellulosic biofuels and chemicals is driven by the increasing global consumption and depletion of fossil resources (Kemppainen 2015). Among the variety of possibilities from glucose-accessible microbial and chemical products, lactic acid, succinic acid and levulinic acid are particularly favourable intermediates for the generation of industrially relevant product family trees.

However, the biorefinery concept that relies on terrestrial crops is under hot debate due to impacts on economy as well as its competition with energy, water and land for food/feed production (Cesário et al. 2018). Therefore, the use of marine resources (e.g. macro- and microalgal biomass) and agri-food residues is attracting the attention of researchers (Cesário et al. 2018). The former is known as a sustainable source of simple sugars commonly destined to bioethanol fermentation. However, alternative uses are receiving attention in the literature in recent years regarding the saccharification and hydrolysation of its sugars since they are carbohydrate-rich feedstocks (Cesário et al. 2018). Levulinic acid, lactic acid, citric acid, polyhydroxyalkanoates, 2,3-butanediol or even succinic acid could be obtained from algal biomass (Ramesh and Kalaiselvam 2011; Hwang et al. 2012; Mazumdar et al. 2013; Alvarado-Morales et al. 2015; Alkotaini et al. 2016; Marinho et al. 2016). The high growth rates associated with its high photosynthetic rate (Jung et al. 2013), the absence of lignin and low presence of hemicellulose support the interest on this type of raw material. Nevertheless, efforts should be conducted in terms of improving the existing lignocellulosic-based technologies (hydrolysis and fermentation) adapted to algal biomass due to the type of carbohydrates (e.g. alginate). Moreover, genetic transformation is also performed to increase the carbohydrate content (Mikami 2013).

Regarding the use of agri-food residues, their exploitation is speculated to increase in the future mainly by emerging technologies not only on second-generation biofuels production but also on high-added value products recovery (Wiloso et al. 2014; Gullón et al. 2018). In 2004, the US Department of Energy identified the 'top' chemicals (up to 50) that could be obtained from biorefinery carbohydrates considering lignocellulosic biomass. In 2009, Bozell and Petersen (2010) reduced the list into ten chemicals which are ethanol, furans, glycerol (and derivatives), biohydrocarbons, lactic acid, succinic acid, hydroxypropionic acid, levulinic acid, sorbitol and xylitol. From that list, only ethanol, furfural, glycerol and sorbitol are available at commercial scale from a biochemical approach.

Organic acids (i.e. lactic acid, succinic acid, hydroxypropionic acid and levulinic acid) are produced in a minimum number of steps from biorefinery carbohydrate streams and have drawn much attention by industrial biotechnology due to their multiple applications as well as they are precursors of several industrially valuable products (Chun et al. 2014). A demanding challenge for the biological production of organic acids at commercially meaningful high titters is to deal with the toxic effects of those acids on the cell growth and cellular metabolisms of the microorganisms producing the acids (Chun et al. 2014).

Although lactic acid has several uses, some of the most expanding ones are the production of biodegradable plastics and textile fibres by means of its polymerisation into polylactic acid (Ilmen et al. 2007) as well as platform chemical for the production of green solvents by means of its esterification into lactate esters (Aparicio and Alcalde 2009). Succinic acid or butanedioic acid is considered a building block obtained by the biochemical transformation of biorefinery sugars (Pinazo et al. 2015; Zhang et al. 2017). Succinic acid is an outstanding renewable platform chemical mostly due to its functionality and valuable derivatives (López-Garzón et al. 2014). Succinic acid is a precursor of well-known petrochemical products such as 1,4-butanediol, tetrahydrofuran, γ-butyrolactone and polybutylene succinates among others. Moreover, succinic acid presents multiple industrial applications in biodegradable polymers (polyesters, polyamides and polyesteramides), foods (e.g. acidulant, flavorant and sweetener), fine chemicals and pharmaceuticals (Sauer et al. 2008; Pateraki et al. 2016). Regarding hydroxypropionic acid, it is obtained from 3-hydroxypropionaldehyde. Its catalytic dehydration produces acrylic acid, acrylate esters and other commodity chemicals (van Maris et al. 2004). Finally, levulinic acid is of interest as a building block and platform chemical due to its simple and high production yield from the hydrolysis of some types of saccharides such as glucose or fructose (Fitzpatrick 2006; Muranaka et al. 2014). Among its potential uses, it can be used to make materials such as plastics and rubbers as well as intermediate for medical supplies.

3 Materials and Methods

3.1 Description of Environmental Assessment

LCA is considered one of the most developed tools for looking holistically at the environmental consequences linked to the life cycle of production processes, products or services. In this sense, it is widely used by environmental professionals and policymakers for the systematic evaluation of the environmental dimension of sustainability. Numerous studies focused on chemical processes have been environmentally assessed following the ISO 14040 (2006) guidelines (Kralisch et al. 2014; Al-Salem et al. 2014). In addition, several authors have explored the implementation of LCA methodology in environmental studies of biorefineries (Neupane et al. 2013; Gilani and Stuart 2015; González-García et al. 2016, 2018) being bio-succinic acid also analysed (Moussa et al. 2016; Smidt et al. 2016). Therefore, its applicability in this area is justified. In this chapter, LCA methodology has been followed in detail considering the specification reported in the ISO standards (ISO 14040 2006).

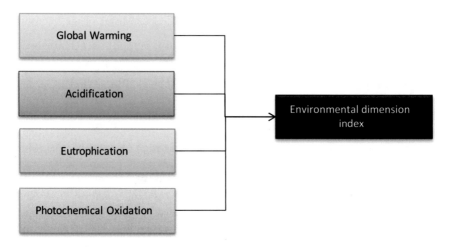

Fig. 2 List of categories selected for assessing the environmental dimension

Among the steps defined within the Life Cycle Impact Assessment (LCIA) of the standardised LCA tool (ISO 14040 2006), classification and characterisation steps have been followed in this study to analyse the production of the bioproducts under assessment from an environmental approach.

The characterisation factors reported by the Centre of Environmental Science of Leiden University—CML 2001 method v2.05 (Guinée et al. 2001) have been considered in this study for the analysis. The following impact categories have been evaluated (see Fig. 2): global warming potential (GWP), acidification potential (AP), eutrophication potential (EP) and photochemical oxidation potential (POP). The choice of these impact categories is that all together gives a complete and comprehensive overview of the environmental effects related to the bioprocesses under evaluation.

Finally, normalisation factors established by the mentioned method (Guinée et al. 2001) have been considered in order to obtain an environmental dimension index per bioproduct. However, since the bioproducts under study are not substitutes, it does not make sense the comparison between their environmental profiles.

The SimaPro v8.2 (PRé Consultants 2017) software has been managed for the computational implementation of the Life Cycle Inventory data in all the case studies (Goedkoop et al. 2013).

3.2 Description of Social Assessment

A social life-cycle assessment (S-LCA) is a method that can be used to assess the social and/or sociological aspects of products or services along the life cycle. Since this method follows a life-cycle assessment approach, it looks at the extraction and processing of raw materials, manufacturing, distribution, use, reuse, maintenance, recycling and final disposal (UNEP-SETAC 2009).

S-LCA makes use of generic and site-specific data or indicators, which can be quantitative, semi-quantitative or qualitative such as annual salary, working hours per week, forced labour, discrimination, child labour, women-to-men ratio of employees or number of accidents. Quantitative indicators describe the analysed issue based on numbers, whereas qualitative indicators describe an issue using words. Finally, semi-quantitative indicators categorise qualitative indicators into a 'YES/NO' form or a scoring system.

Therefore, it perfectly complements the social dimension in sustainability assessments. The UNEP-SETAC guidelines recommend a similar method within the framework of the ISO 14040 (2006), including the four stages of goal and scope definition, life cycle inventory, life cycle impact assessment and interpretation of results (UNEP-SETAC 2009). Thus, the guidelines describe social impacts as '*consequences of positive or negative pressures on social endpoints*' (UNEP-SETAC 2009).

Treatment of social well-being is relatively new in the field of quantitative impact assessment at product and technology level. Thus, the measurement of the impact on social well-being with a life cycle perspective needs special attention. The social impact assessment should include impacts on human well-being which include a broad range of pathways that affect the quality of life of people (Weidema 2006) such as (i) autonomy of people—directly related with forced labour, (ii) safety and security—associated with unemployment and (iii) equality—negatively impacted by income distribution, fair salary or equal opportunities.

There are many possible indicators to address impacts on social dimension and, in this study, eight main indicators have been selected as most relevant for the bioproducts under study and taking into account the availability of required data. These indicators are displayed in Fig. 3 classified in terms of their potential contribution to the different social pathways that affect the quality of life and considering the different stakeholder categories mentioned by UNEP-SETAC (2009) to which they can contribute.

Finally, a social dimension index (see Fig. 3) has been estimated taking into account an equal contribution from the three social pathways, tranquility and security correspond with a type of social indicators category described below which takes into account aspects such as transparecy, working hours, health of people involved in byproducts production (autonomy, safety, security and tranquility and, equality).

Operationalisation of indicators

Assessing in more detail the eight indicators chosen, the following sub-categories within indicator and social pathway should be required (if possible) from stakeholders:

Safety, security and tranquillity

- Working hours: total working hours per month or year in bioproducts-related plants. This issue should be really interesting to be considered in the analysis, however, it has not been included due to lack of good quality data.

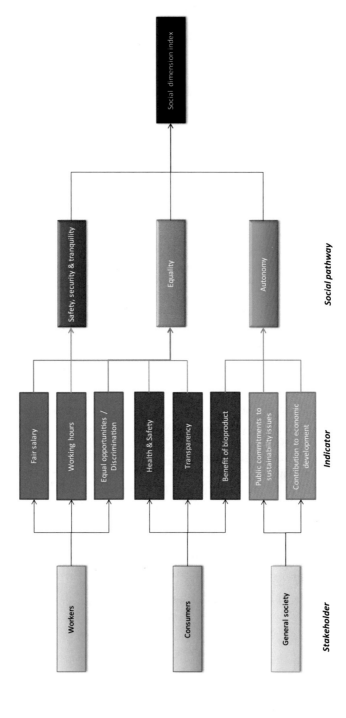

Fig. 3 List of stakeholders, indicators and social pathways selected to assess the social dimension

- Health and safety: information regarding the performance of safety tests, availability of safety test for checking, regulations…
- Transparency: information regarding the availability of consumer service, consumers' complaints files, supply of information to the consumers regarding characteristics of the bioproduct.

Equality

- Fair salary: annual salary of bioproducts workers as well as identification of differences in the salary between men and women.
- Equal opportunities/discrimination: identification of women-to-men differences not only in the salary for similar work but also in the labour force participation.

Autonomy

- Benefits of bioproduct: information regarding the perception of added value of the bioproduct in comparison with fossil alternatives, valuation of the use of agri-food wastes as raw material instead of other sources.
- Public commitments to sustainability issues: public perception of environmental benefits linked to the bioproduct in comparison with fossil one, public perception of agri-food wastes use to bioproducts obtaining.
- Contributions to economic development: public perception of bioproducts, public perception of agri-food wastes use to bioproducts obtaining, potential market of bioproducts.

3.3 Case Studies

3.3.1 Oligosaccharides and Antioxidants Production from Agri-food Industry Waste

In the last decade, consumers have a greater concern for health, which has driven the demand for novel functional foods enriched with bioactive compounds (Grajek et al. 2005). These bioactive compounds can be obtained from low-cost agricultural and agro-industrial by-products using environmentally friendly technologies (Gullón et al. 2014).

In this chapter, it has been considered the obtaining of high-added value products from vine shoots. This waste from the wine sector has been traditionally poorly exploited, however, various studies focused on its composition suggests presence of a wide range of bioactive compounds such as oligosaccharides and antioxidant compounds with potential nutraceuticals and pharmaceutical applications (Dávila et al. 2016; Gullón et al. 2017).

The valorisation of this residue into high-added value products has been performed considering the three steps which are described below:

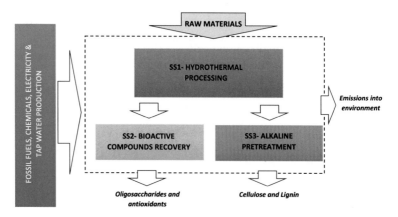

Fig. 4 System boundaries and stages involved in vine shoots-based oligosaccharides and antioxidants production system

First, vine shoots are subjected to a hydrothermal processing (SS1) to separate its main structural components (hemicellulosic oligosaccharides and antioxidants compounds in liquid stream and a solid fraction rich in cellulose and lignin).

The liquid phase is processed to obtain two different streams of bioactive compounds, one containing oligosaccharides and the other one with antioxidants. For this purpose, this liquid phase is extracted with ethyl acetate, and aqueous and organic phases are separated by decantation. The organic phase was vacuum evaporated to obtain an extract rich in antioxidant compounds and aqueous phase is concentrated using membranes to obtain other streams rich in oligosaccharides. These stages are part of the subsystem 'Bioactive Compounds Recovery'.

Finally, the solid phase is subjected to alkaline delignification to recovery separately the cellulose and lignin. This stage is based on a mild thermal treatment using NaOH to solubilise the lignin and provides a solid-phase enriched into cellulose. Several steps of filtration and washing with water are involved in this stage.

A summarised scheme of the production of high-added value compounds considered in this study is displayed in Fig. 4.

3.3.2 Succinic Acid Production from Agri-food Industry Waste

Succinic acid or butanedioic acid ($C_4H_6O_4$) is an outstanding renewable platform chemical supported by its functionality and valuable derivatives (López-Garzón et al. 2014). There is a wide recent literature focused on its production and interest as chemical building block (Zhang et al. 2017; Moussa et al. 2016; Smidt et al. 2016; Pinazo et al. 2015; Li et al. 2010). It can be obtained from biochemical transformation of biorefinery sugars (bacterial fermentation of carbohydrates), from a range of feedstocks and considering multiple microorganisms (Orjuela et al. 2013). In addition, environmental benefits are linked to its bioproduction since carbon dioxide is needed by microorganisms (carbon dioxide fixation involved in the reductive TCA cycle) (Pateraki et al. 2016; Bechtold et al. 2008).

In this chapter, it has been considered the production of succinic acid from apple pomace as raw material, which is considered as a residue in apple and juice industries.

The valorisation of apple pomace into succinic acid has been performed considering the three main steps, which are summarised below:

First, the apple pomace is received from the juice factory and it is warehoused in hoppers. Next, it is dried at atmospheric pressure in a tray drier (60 °C) with the aim of reducing its moisture content and increasing its lifespan. The dried raw material is stored in silos at 20 °C and atmospheric pressure to guarantee its conservation and to avoid the proliferation of plagues. These activities are involved in the *Raw material reconditioning and storage stage* (SS1).

Second, the sugars fermentation takes place under a simultaneous saccharification and fermentation—SSF step (*SSF stage*, SS2). As a difference to other valorisation routes, the microorganism used in this process (*A. succinogenes*) requires the consumption of carbon dioxide and glucose as carbon sources.

Finally, the purification of the succinic acid (*Purification stage*, SS3) is performed in order to obtain succinic acid at industrial grade (pure A-grade, i.e. ≥ 99.5 wt%). Multiple activities are involved in this stage such as membranes ultrafiltration, reactive extraction with tri-n-octilamine (TOA) in 1-octanol and vacuum distillation (and crystallisation) to obtain the desired pure succinic acid stream. A summarised scheme of the succinic acid production sequence under study is depicted in Fig. 5.

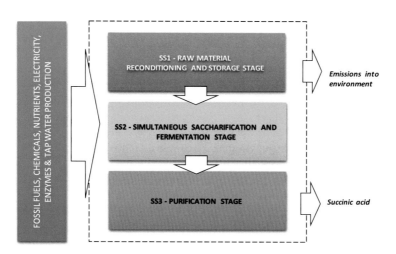

Fig. 5 System boundaries and stages involved in apple pomace-based succinic acid production system

3.4   Life Cycle Inventory Data Acquisition

A reliable environmental assessment requires the collection of high-quality inventory data. A consistent environmental assessment requires the collection of high-value life cycle inventory (LCI) data.

In this study, inventory data for the foreground systems (i.e. direct inputs and outputs for each process or step) correspond with average data taken from the modellisation at full scale of each biorefinery scenario. To do so, information from the laboratory has been used to design and model the production sequences.

Primary data correspond to electricity requirements in the different units: reactors, centrifuges, membranes, orbital shakers, distillers, freeze dryers, etc., as well as to the use of chemicals, enzymes, nutrients and tap water, depending on the valorisation route. Secondary data have been also managed but only for the background processes such as production of electricity, chemicals, nutrients and tap water, which have been taken from the Ecoinvent database® v3.1 (Wernet et al. 2016). Regarding enzymes production, the inventory data have been taken from Gilpin and Andrae (2017).

Ancillary activities such as wastewater and solid waste treatment have been also included within the system boundaries in order to compute the environmental impacts from the different wastes management. Inventory data corresponding to wastewater treatment activities has been taken from Doka (2007). Regarding solid wastes, it has been assumed their management in sanitary landfills (Doka 2007).

4   Results and Discussion

4.1   Environmental Sustainability of Bioproducts

4.1.1   Oligosaccharides and Antioxidants Production from Agri-food Industry Waste

Figure 6 displays the characterisation results per kg of valorised agri-food waste as well as the environmental dimension index associated to the production system. It is important to bear in mind that 170 g of hemicellulosic oligosaccharides and 27 g of antioxidant extract are obtained per kg of vine shoots valorised.

If it is taken a look at the environmental profile displayed in Fig. 6, it is possible to identify the stages responsible for the highest impacts. Figure 7 identifies two main responsible stages that are the bioactive compounds recovery (SS2) and the alkaline pretreatment (SS3).

The bioactive compounds recovery stage includes the hemicellulosic oligosaccharides extraction with ethyl acetate, the further extractive chemical recovery with the production of the antioxidants extract and the final oligosaccharides freeze drying (Gullón et al. 2018). Production of electricity requirements in the freeze drying and in the vacuum evaporation to recover the ethyl acetate and to obtain the

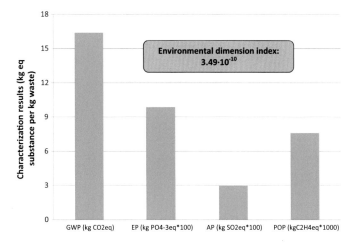

Fig. 6 Environmental profile associated with the production of both oligosaccharides and antioxidants from vine shoots

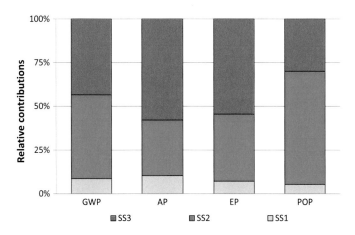

Fig. 7 Distribution of impacts per stages involved in the valorisation sequence

extract is considered as the main responsibility of environmental burdens derived from this stage with contributing ratios higher than 98% of impacts from SS2.

The alkaline pretreatment stage manages the solid fraction from SS1, where it is subjected to alkaline delignification and different filtration and precipitation steps to obtain lignin and cellulose as co-products. Production of electricity requirements in the delignification process is also considered as an environmental hotspot together with the production of the sulfuric acid which is required to precipitate the lignin from the black liquor. Contributions from the former are 96, 86, 75 and 86% of GWP, AP, EP and POP of total SS3. Contributions from the latter are 2, 12, 55 and 12% of GWP, AP, EP and POP of total SS3.

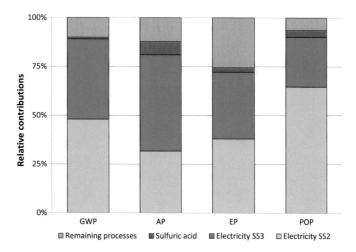

Fig. 8 Identification of environmental hotspots

Thus, electricity and sulfuric acid production plays a key environmental role (see Fig. 8) and environmental improvements should be focused on optimising their requirements.

4.1.2 Succinic Acid Production from Food Industry Waste

Figure 9 displays the characterisation results per kg of valorised waste as well as the environmental dimension index associated with the production system. It is important to bear in mind that 0.55 kg of succinic acid is obtained per kg of apple pomace valorised.

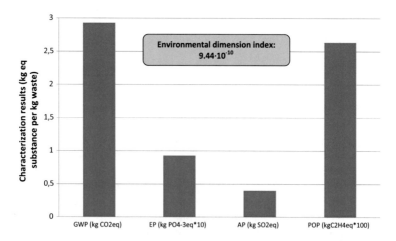

Fig. 9 Environmental profile associated with the production of succinic acid from apple waste

Having in mind the contributions from the different stages involved in the production system to the global environmental profile, the purification stage is the environmental hotspot since it is responsible for contributions higher than 93% in all the categories considered for analysis. The rationale behind the large environmental burdens derived from this stage is associated with the use of organic chemicals required in the extraction process as well as the large electricity requirements mostly in the distillation process. Although 95% of total organic chemicals dose required in the extraction is recovered in the distillation process and recycled to the system, the effect from chemicals waste into the environmental profile is outstanding.

Production of wasted chemicals ($\sim 5\%$ of total dose) is responsible for 13, 37, 37 and 64% of total contributions to GWP, AP, EP and POP, respectively. Contributions from the production of electricity demand in the distillation process are 84, 56, 46 and 31% of total GWP, AP, PE and POP, respectively. Thus, further optimisation activities should be focused on both processes to reduce electricity demand, and to identify alternative chemicals with a better environmental background or to increase the recovery ratio in the distillation process. As difference to other valorisation strategies reported in the literature for alternative agri-food wastes where pretreatment stage considerably affects the environmental profiles due to the consideration of autohydrolysis (González-García et al. 2016, 2018; Gullón et al. 2018), in the succinic acid production system under study, it is not required a specific pretreatment process due to the characteristics of the raw material considered for valorisation (apple pomace) since it presents a huge amount of soluble sugars and a high susceptibility to the enzymatic hydrolysis.

Others methods for the purification of succinic acid have been suggested in the literature including electrodialysis, precipitation with ammonia or calcium hydroxide, pre-dispersed solvent extraction with colloidal liquid aphrons or ion exchange as an alternative to the use of organic solvents (Kurzrock and Weuster-Botz 2010). The analysis of their effect on the environmental profile derived from succinic acid production should be analysed to identify more attractive production routes from an environmental approach.

It is important to bear in mind that the environmental burdens linked to the production of the enzymes dose required in SS2 have not been taken into consideration in the profile reported in Fig. 9. The rationale behind this decision is that enzymes production process (background process) has not been included in other studies available in the literature focused on butanedioic acid production (Moussa et al. 2016; Smidt et al. 2016). Thus, its consideration should complicate a direct comparison between environmental results. However and according to our results, the enzymes production process should be considered since it is a high-energy intensive process (Gilpin and Andrae 2017). Therefore, including the production of enzymes dose in the environmental profile, the environmental burdens should considerably be increased being 95 times higher in terms of GWP, 4 times higher in the remaining categories. Thus, special attention must be paid into the enzymes production process considered for analysis since it plays an environmental key role.

4.2 Social Sustainability of Bioproducts

One of the main challenges of this chapter is the assessment of social dimension of both bioproducts production chains considered for analysis. Since the social dimension can be assessed considering different indicators or strategies, the profiles have been analysed taking into account the social indicators reported in Fig. 3.

First, social information must be gathered. Thus, specific questionnaires were designed and supplied to different related stakeholders (workers, consumers and general society). It is important to bear in mind that both scenarios proposed for analysed are carried out at pilot scale or simulated. Thus, stakeholders related with similar bioproducts manufacturing were interviewed. Information related with stakeholders is not reported in this chapter due to confidential issues.

Aggregation and quantification of indicators

The sub-categories selected for analysis are qualitative (i.e. Yes/No presentation) or semi-quantitative. Thus, the qualitative ones were converted into semi-quantitative by means of a scoring strategy being the mark 1 allocated to the 'No' answer and the mark 3 to the 'Yes' answer. Regarding the semi-quantitative ones, the scores 1 (worse value)—2–3 (best value) were marked by the stakeholders according to their feelings and knowledge regarding the topic. According to it, it was possible to obtain a quantitative final score to each sub-category and indicator as well as to estimate the social dimension index.

All the indicators were aggregated in order to come to one overall quantitative score for the impact on the social dimension. It is important to remark here that each pathway of social well-being considered (safety, security and tranquillity, equality and autonomy) contributes equality (1/3) to the estimation of the final social score.

The aggregated score for each indicator (I) was calculated as the arithmetical mean of the weighted value of involved sub-categories (Sc) taking into account the sample of stakeholders, the scores marked by each stakeholder and the number of sub-categories per indicator (Eqs. 1 and 2). Next, the social index (S) is estimated considering the results obtained in each social pathway (Eq. 3) by means of an arithmetical mean (Eq. 4) and taking into account the number of indicators considered per pathway.

$$Sc_y = \sum_{i=1}^{n} v_i \cdot \frac{1}{n} \tag{1}$$

$$I_z = \sum_{y=1}^{j} Sc_y \cdot \frac{1}{j} \tag{2}$$

$$P_x = \sum_{z=1}^{t} I_z \cdot \frac{1}{t} \tag{3}$$

$$S = \sum_{x=1}^{3} P_x \cdot \frac{1}{3} \qquad (4)$$

where v_i is the score marked by the stakeholder i, Sc_y is the average value of a sub-category y, I_z is the average value of an indicator z, P_x is the value corresponding to the social pathway x, S is the Social dimension score, n is the number of stakeholders that constitute the sample, j is the number of sub-categories that constitute an indicator, t is the number of indicators that constitute a social pathway and x is the number of social pathways.

Social dimension results

According to the questionnaires supplied by the consulted stakeholders, the results obtained per scenario under assessment taking into account the social issues are displayed in Fig. 10.

Social indicators such as Health and safety, Transparency, Fair salary and Contributions to economic development report higher scores for the prebiotics and antioxidant extracts production than for succinic acid production. On the contrary, the succinic acid production derives into a better profile in terms of Equal opportunities/discrimination and Benefits of bioproduct. The consideration of succinic acid of a 'top chemical' could be associated with the social perception regarding this bioproduct since the other two are related with food and pharmaceutical uses. However, it must be highlighted that not so outstanding differences are identified in Public dedication to sustainability issues ($\sim$3%).

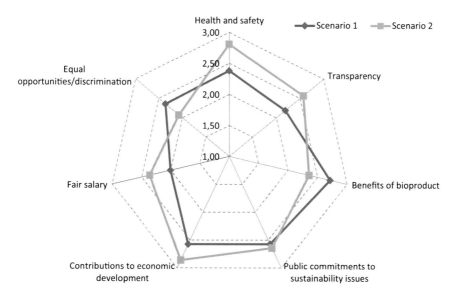

Fig. 10 Social results per social indicators considered for analysis. Scenario 1—Succinic acid production; Scenario 2—Oligosaccharides and antioxidants production

Regarding the social results per assessed stakeholder, Fig. 11 displays the comparative scores and regardless of the stakeholder, the production of oligosaccharides and antioxidant is better scored. The consideration of oligosaccharides as prebiotics and their nutraceutical and pharmaceutical properties are behind the results obtained from consumers and general society. Regarding the scores obtained from workers, they are very similar since there are no remarkable differences between manufacturing stakeholders consulted.

According to Fig. 12, the production of oligosaccharides and antioxidant extract from agri-waste reports a better social perception (6% higher) than the succinic acid-based scenario. The rationale behind these results could be linked to the characteristics of the bioproducts since prebiotics belongs to pharmaceutical and nutraceutical sectors and, nowadays, the society is considerably concerned by health and overall quality of life, demanding novel functional products with the ability to prevent diseases and to maintain the human health.

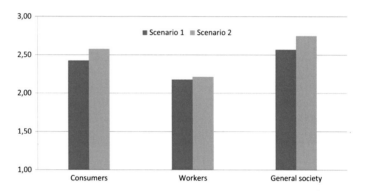

Fig. 11 Comparative social scores per stakeholder

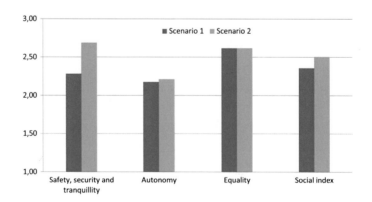

Fig. 12 Social results per social pathways and global social index (S). Scenario 1—Succinic acid production; Scenario 2—Oligosaccharides and antioxidants production

5 Conclusions and Future Outlook

Nowadays, special attention is being paid to bioproducts obtaining from agri-food residues not only due to their potential uses as alternative compounds to conventional fossil ones but also due to the valorisation of an organic waste from circular economy and biorefinery approaches.

Environmental benefits have been assessed in this chapter as well as in the literature. However, other issues need to be addressed to demonstrate their sustainability regardless of conventional products. Therefore, it is imperative to have tools or methodologies for sustainability assessment taking into account social and economic pillars.

Thus, the environmental pillar has been complemented in this chapter with the social one to analyse two different biorefinery scenarios focused on the valorisation of different agri-food wastes, i.e. vine shoots and apple pomace. In addition, a methodology to analyse the social dimension has been formulated considering multiple social sub-categories related with social well-being.

Nevertheless and in order to obtain a final sustainability index, further research is required focused on the assessment of the economic pillar.

Acknowledgements This research has been financially supported by Xunta de Galicia (project ref. ED431F 2016/001), the Spanish Ministry of Economy and Competitiveness (CTQ2016-81848-REDT) and the STAR-ProBio project funded by the European Union's Horizon 2020 Program (Grant agreement No. 727740). S.G.-G., P.G. & B.G. would like to express their gratitude to the Spanish Ministry of Economy and Competitiveness for financial support (Grant references RYC-2014-14984, IJCI-2015-25304 and IJCI-2015-25305, respectively). The authors S.G.-G. & B.G. belong to the Galician Competitive Research Group GRC 2013-032, programme co-funded by FEDER as well as to CRETUS (AGRUP2015/02).

References

Agócs, A., Nagy, V., Szabó, Z., Márk, L., Ohmacht, R., & Deli, J. (2007). Comparative study on the carotenoid composition of the peel and the pulp of different citrus species. *Innovative Food Science and Emerging Technologies, 8,* 390–394.

Alañón, M. E., Alarcón, M., Marchante, L., Díaz-Maroto, M. C., & Pérez-Coello, M. S. (2017). Extraction of natural flavorings with antioxidant capacity from cooperage by-products by green extraction procedure with subcritical fluids. *Industrial Crops and Products, 103,* 222–232.

Alkotaini, B., Koo, H., & Kim, B. S. (2016). Production of polyhydroxyalkanoates by batch and fed-batch cultivations of *Bacillus megaterium* from acid-treated red algae. *Korean Journal of Chemical Engineering, 33,* 1669–1673.

Al-Salem, S. M., Evangelisti, S., & Lettieri, P. (2014). Life cycle assessment of alternative technologies for municipal solid waste and plastic solid waste management in the Greater London area. *Chemical Engineering Journal, 244,* 391–402.

Alvarado-Morales, M., Gunnarsson, I. B., Fotidis, I. A., Vasilakou, E., Lyberatos, G., & Angelidaki, I. (2015). *Laminaria digitata* as a potential carbon source for succinic acid and bioenergy production in a biorefinery perspective. *Algal Research, 9,* 126–132.

Aparicio, S., & Alcalde, R. (2009). The green solvent ethyl lactate: An experimental and theoretical characterization. *Green Chemistry, 11,* 65–78.

Ayala-Zavala, J. F., Vega-Vega, V., Rosas-Domínguez, C., Palafox-Carlos, H., Villa-Rodriguez, J. A., Wasim Siddiqui, Md, et al. (2011). Agro-industrial potential of exotic fruit byproducts as a source of food additives. *Food Research International, 44,* 1866–1874.

Banerjee, J., Singh, R., Vijayaraghavan, R., MacFarlane, D., Patti, A. F., & Arora, A. (2017). Bioactives from fruit processing wastes: Green approaches to valuable Chemicals. *Food Chemistry, 225,* 10–22.

Barba, F. J., Zhu, Z., Koubaa, M., Sant'Ana, A. S., & Orlien, V. (2016). Green alternative methods for the extraction of antioxidant bioactive compounds from winery wastes and by-products: A review. *Trends in Food Science & Technology, 49,* 96–109.

Baysal, T., Ersus, S., & Starmans, D. A. (2000). Supercritical (CO_2) extraction of β-carotene and lycopene from tomato paste waste. *Journal of Agricultural and Food Chemistry, 48,* 5507–5511.

Bechtold, I., Bretz, K., Kabasci, S., Kopitzky, R., & Springer, A. (2008). Succinic acid: A newplatform chemical for biobased polymers from renewable resources. *Chemical Engineering Technology, 31,* 647–654.

Blaga, S. (2013). *Corporate social responsibility from a Romanian perspective.* Cluj-Napoca, Romania: University Babes-Bolay Press. ISBN 978-973-595-540-3.

Blok, K., Huijbregts, M., Patel, M.K., Hertwich, E., Hauschild, M., Sellke, P., et al. (2013). Utrecht University Repository (Book). Report prepared within the EC 7th Framework (Project n°: 227078), Project title: Development and application of a standardized methodology for the PROspective SUstaInability Assessment of Technologies.

Bond, A. J., & Morrison-Saunders, A. (2011). Re-evaluating sustainability assessment: Aligning the vision and the practice. *Environmental Impact Assessment Review, 31,* 1–7.

Bozell, J. J., & Petersen, G. R. (2010). Technology development for the production of biobased products from biorefinery carbohydrates-the US Department of Energy's "Top10" revisited. *Green Chemistry, 12,* 539–554.

Budzianowski, W. M., & Postawa, K. (2016). Total chain integration of sustainable biorefinery systems. *Applied Energy, 184,* 1432–1446.

Cesário, M. T., da Fonseca, M., Manuela, R., Marques, M. M., de Almeida, M., & Catarina, M. D. (2018). Marine algal carbohydrates as carbon sources for the production of biochemicals and biomaterials. *Biotechnology Advances, 36,* 798–817.

Chun, Y., Yunxiao, L., Ashok, S., Seol, E., & Park, S. (2014). Elucidation of toxicity of organic acids inhibiting growth of *Escherichia coli* W. *Biotechnology and Bioprocess Engineering, 19,* 858–865.

Dávila, D., Gordobil, O., Labidi, J., & Gullón, P. (2016). Assessment of suitability of vine shoots for hemicellulosic oligosaccharides production through aqueous processing. *Bioresource Technology, 211,* 636–644.

Deng, G. F., Shen, C., Xu, X. R., Kuang, R. D., Guo, Y. J., Zeng, L. S., et al. (2012). Potential of fruit wastes as natural resources of bioactive compounds. *International Journal of Molecular Sciences, 13,* 8308–8323.

Doka, G. (2007). Life cycle inventories of waste treatment services. Ecoinvent Report No 13, Dübendorf, Switzerland.

dos Santos, T. C., Gomes, D. P. P., Bonomo, R. C. F., & Franco, M. (2012). Optimisation of solid state fermentation of potato peel for the production of cellulolytic enzymes. *Food Chemistry, 133,* 1299–1304.

Fava, F., Totaro, G., Diels, L., Reis, M., Duarte, J., Beserra Carioca, O., et al. (2015). Biowaste biorefinery in Europe: Opportunities and research & development needs. *New Biotechnology, 32,* 100–108.

Fernández-Bolaños, J., Rodríguez, G., Gómez, E., Guillén, R., Jiménez, A., Heredia, A., et al. (2004). Total recovery of the waste of two-phase olive oil processing: Isolation of added-value compounds. *Journal of Agricultural and Food Chemistry, 52,* 5849–5855.

Fitzpatrick, S. W. (2006). The biofine technology: A "bio-refinery" concept based on thermochemical conversion of cellulosic biomass. *ACS Symposium Series, 921,* 271–287.

Galanakis, C. M. (2012). Recovery of high added-value components from food wastes: Conventional, emerging technologies and commercialized applications. *Trends in Food Science & Technology, 26,* 68–87.

Gilani, B., & Stuart, P. R. (2015). Life cycle assessment of an integrated forest biorefinery: Hot water extraction process case study. *Biofuels, Bioproducts and Biorefining, 9,* 677–695.

Gilpin, G., & Andrae, A. S. G. (2017). Comparative attributional life cycle assessment of European cellulase enzyme production for use in second-generation lignocellulosic bioethanol production. *The International Journal of Life Cycle Assessment, 22,* 1034–1053.

Goedkoop, M., Oele, M., Leijting, J., Ponsioen, T., & Meijer, E. (2013). *Introduction to LCA with SimaPro 8.* The Netherlands: PRé Consultants.

Gómez, B., Gullón, B., Yañez, R., Parajó, J. C., & Alonso, J. L. (2013). Pectic-oligosacharides from lemon peel wastes: Production, purification and chemical characterization. *Journal of Agricultural and Food Chemistry, 61,* 10043–10053.

González-García, S., Gullón, B., & Moreira, M. T. (2018). Environmental assessment of biorefinery processes for the valorization of lignocellulosic wastes into oligosaccharides. *Journal of Cleaner Production, 172,* 4066–4073.

González-García, S., Gullón, B., Rivas, S., Feijoo, G., & Moreira, M. T. (2016). Environmental performance of biomass refining into high-added value compounds. *Journal of Cleaner Production, 120,* 170–180.

González-Paramás, A. M., Esteban-Ruano, S., Santos-Buelga, C., de Pascual-Teresa, S., & Rivas-Gonzalo, J. C. (2004). Flavanol content and antioxidant activity in winery byproducts. *Journal of Agricultural and Food Chemistry, 52,* 234–238.

Grajek, W., Olejnik, A., & Sip, A. (2005). Probiotics: Prebiotics and antioxidants as functional foods. *Acta Biochimica Polonica, 52,* 665–671.

Griggs, D., Stafford-Smith, M., Gaffney, O., Rockström, J., Öhman, M. C., Shyamsundar, P., et al. (2013). Policy: Sustainable development goals for people and planet. *Nature, 495,* 305–307.

Guinée, J. B., Gorrée, M., Heijungs, R., Huppes, G., Kleijn, R., de Koning, A., et al. (2001). *Life cycle assessment. An operational guide to the ISO Standards.* Leiden, The Netherlands: Centre of Environmental Science.

Gullón, B., Eibes, G., Moreira, M. T., Dávila, I., Labidi, J., & Gullón, P. (2017). Antioxidant and antimicrobial activities of extracts obtained from the refining of autohydrolysis liquors of vine shoots. *Industrial Crops and Products, 107,* 105–113.

Gullón, P., González-Muñoz, M. J., van Gool, M. P., Schols, H. A., Hirsch, J., Ebringerová, A., et al. (2010). Production, refining, structural characterization and fermentability of rice husk xylooligosaccharides. *Journal of Agricultural and Food Chemistry, 58,* 3632–3641.

Gullón, P., Gullón, B., Dávila, I., Labidi, J., & Gonzalez-Garcia, S. (2018). Comparative environmental life cycle assessment of integral revalorization of vine shoots from a biorefinery perspective. *Science of the Total Environment, 624,* 225–240.

Gullón, B., Gullón, P., Sanz, Y., Alonso, J. L., & Parajó, J. C. (2011). Prebiotic potential of a refined product containing pectic oligosaccharides. *LWT—Food Science and Technology, 44,* 1687–1696.

Gullón, B., Gullón, P., Tavaria, F., Pintado, M., Gomes, A. M., Alonso, J. L., et al. (2014). Structural features and assessment of prebiotic activity of refined arabinoxylooligosaccharides from wheat bran. *Journal of Functional Foods, 6,* 438–449.

Hollmann, J., & Lindhauer, M. G. (2005). Pilot-scale isolation of glucuronoarabinoxylans from wheat bran. *Carbohydrate Polymers, 59,* 225–230.

Hwang, H. J., Kim, S. M., Chang, J. H., & Lee, S. B. (2012). Lactic acid production from seaweed hydrolysate of *Enteromorpha prolifera* (chlorophyta). *Journal of Applied Phycology, 24,* 935–940.

Ilmen, M., Koivuranta, K., Ruohonen, L., Suominen, P., & Penttila, M. (2007). Efficient production of L-lactic acid from xylose by *Pichia stipitis*. *Applied and Environmental Microbiology, 73,* 117–123.

Iribarren, D., Martín-Gamboa, M., O'Mahony, T., & Dufour, J. (2016). Screening of socio-economic indicators for sustainability assessment: A combined life cycle assessment and data envelopment analysis approach. *The International Journal of Life Cycle Assessment, 21,* 202–214.

ISO 14044 (2006). *Environmental management e life cycle assessment e principles and framework.* Geneva: International Organization for Standardization.

ISO 14045. (2012). *Environmental management-eco-efficiency assessment of product systems-principles, requirements and guidelines.* Geneva: International Organization for Standardization.

Jørgensen, A., Le Bocq, A., Nazarkina, L., & Hauschild, M. (2008). Methodologies for social life cycle assessment. *The International Journal of Life Cycle Assessment, 13,* 96–103.

Jung, K. A., Lim, S. R., Kim, Y., & Park, J. M. (2013). Potentials of macroalgae as feedstocks for biorefinery. *Bioresource Technology, 135,* 182–190.

Kam, B. (2009). Carbohydrate-based food processing wastes as biomass for biorefining of biofuels and chemicals (chapter 20). In *Food Science, Technology and Nutrition: Volume 2. Handbook of waste management and co-product recovery in food processing* (pp. 479–514). Woodhead Publishing Series.

Kemppainen, K. (2015). *Production of sugars, ethanol and tannin from spruce bark and recovered fibres* (Doctoral thesis). Aalto University, Department of Biotechnology and Chemical Technology, Finland. ISBN 978-951-38-8215-0 (electronic). http://www.vtt.fi/inf/pdf/science/2015/S76.pdf.

Kralisch, D., Ott, D., & Gericke, D. (2014). Rules and benefits of life cycle assessment in green chemical process and synthesis design: A tutorial review. *Green Chemistry, 17,* 123–145.

Kurzrock, T., & Weuster-Botz, D. (2010). Recovery of succinic acid from fermentation broth. *Biotechnololgy Letters, 32,* 331–339.

Lama-Muñoz, A., Romero-García, J. M., Cara, C., Moya, M., & Castro, E. (2014). Low energy-demanding recovery of antioxidants and sugars from olive stones as preliminary steps in the biorefinery context. *Industrial Crops and Products, 60,* 30–38.

Lavelli, V., & Torresanim, M. C. (2011). Modelling the stability of lycopene-rich by-products of tomato processing. *Food Chemistry, 125,* 529–535.

Lenucci, M. S., Durante, M., Anna, M., Dalessandro, G., & Piro, G. (2013). Possible use of the carbohydrates present in tomato pomace and in byproducts of the supercritical carbon dioxide lycopene extraction process as biomass for bioethanol production. *Journal of Agricultural Food Chemistry, 61,* 3683–3692.

Li, X., He, X., Lv, Y., & He, Q. (2014). Extraction and functional properties of water-soluble dietary fiber from apple pomace. *Journal of Food Process Engineering, 37,* 293–298.

Li, Q., Wang, D., Wu, Y., Li, W., Zhang, Y., Xing, J., et al. (2010). One step recovery of succinic acid from fermentation broths by crystallization. *Separation and Purification Technology, 72,* 294–300.

López-Garzón, C. S., van der Wielen, L. A. M., & Straathof, A. J. J. (2014). Green upgrading of succinate using dimethyl carbonate for a better integration with fermentative production. *Chemical Engineering Journal, 235,* 52–60.

Lun, O. K., Wai, T. B., & Ling, L. S. (2014). Pineapple cannery waste as a potential substrate for microbial biotranformation to produce vanillic acid and vanillin. *International Food Research Journal, 21,* 953–958.

Marín, F. R., Soler-Rivas, C., Benavente-García, O., Castillo, J., & Pérez-Alvarez, J. A. (2007). By-products from different citrus processes as a source of customized functional fibres. *Food Chemistry, 100,* 736–741.

Marinho, G. S., Alvarado-Morales, M., & Angelidaki, I. (2016). Valorization of macroalga *Saccharina latissima* as novel feedstock for fermentation-based succinic acid production in a biorefinery approach and economic aspects. *Algal Research, 16,* 102–109.

Martínez, M., Gullón, B., Schols, H. A., Alonso, J. L., & Parajó, J. C. (2009). Assessment of the production of oligomeric compounds from sugar beet pulp. *Industrial and Engineering Chemistry Research, 48,* 4681–4687.

Mazumdar, S., Lee, J., & Oh, M. K. (2013). Microbial production of 2,3 butanediol from seaweed hydrolysate using metabolically engineered *Escherichia coli. Bioresource Technology, 136,* 329–336.

Mikami, K. (2013). Current advances in seaweed transformation. In: G. R. Baptista (Ed.), *An integrated view of the molecular recognition and toxinology—From analytical procedures to biomedical applications* (pp. 324–347). InTech.

Mohdaly, A. A., Sarhan, M. A., Smetanska, I., & Mahmoud, A. (2010). Antioxidant properties of various solvent extracts of potato peel, sugar beet pulp and sesame cake. *Journal of the Science of Food and Agriculture, 90*, 218–226.

Monrad, J. K., Suárez, M., Motilva, M. J., King, J. W., Srinivas, K., & Howarda, L. R. (2014). Extraction of anthocyanins and flavan-3-ols from red grape pomace continuously by coupling hot water extraction with a modified expeller. *Food Research International, 65*, 77–87.

Moreira, D., Gullón, B., Gullón, P., Gomes, A., & Tavaria, F. (2016). Bioactive packaging using antioxidant extracts for the prevention of microbial food-spoilage. *Food and Function, 7*, 3273–3282.

Moussa, H. I., Elkamel, A., & Young, S. B. (2016). Assessing energy performance of bio-based succinic acid production using LCA. *Journal of Cleaner Production, 139*, 761–769.

Müller-Maatsch, J., Bencivenni, M., Caligiani, A., Tedeschi, T., Bruggeman, G., Bosch, M., et al. (2016). Pectin content and composition from different food waste streams. *Food Chemistry, 201*, 37–45.

Muranaka, Y., Suzuki, T., Sawanishi, H., Hasegawa, I., & Mae, K. (2014). Effective production of levulinic acid from biomass through pretreatment using phosphoric acid, hydrochloric acid, or ionic liquid. *Industrial and Engineering Chemistry Research, 53*, 11611–11621.

Mushtaq, Q., Irfan, M., Tabssum, F., & Qazi, J. I. (2017). Potato peels: A potential food waste for amylase production. *Journal of Food Process Engineering, 40*, 12512–12520.

Muthaiyan, A., Hernandez-Hernandez, O., Moreno, F. J., Sanz, M. L., & Ricke, S. C. (2012). Hydrolyzed caseinomacropeptide conjugated galactooligosaccharides support the growth and enhance the bile tolerance in lactobacillus strains. *Journal of Agricultural and Food Chemistry, 60*, 6839–6845.

Nadour, M., Michaud, P., & Moulti-Mati, F. (2012). Antioxidant activities of polyphenols extracted from olive (*Olea europaea*) of chamlal variety. *Applied Biochemistry and Biotechnology, 167*, 1802–1810.

Ndayishimiye, J., & Chun, B. S. (2017). Optimization of carotenoids and antioxidant activity of oils obtained from a co-extraction of citrus (*Yuzu ichandrin*) by-products using supercritical carbon dioxide. *Biomass and Bioenergy, 106*, 1–7.

Neupane, B., Halog, A., & Lilieholm, R. J. (2013). Environmental sustainability of wood-derived ethanol: A life cycle evaluation of resource intensity and emissions in Maine, USA. *Journal of Cleaner Production, 44*, 77–84.

Norris, C. B. (2014). Data for social LCA. *The International Journal of Life Cycle Assessment, 19*, 261–265.

Norris, C. B., Cavan, D. A., & Norris, G. (2012). Identifying social impacts in product supply chains: Overview and application of the social hotspot database. *Sustainability, 4*, 1946–1965.

Orjuela, A., Orjuela, A., Lira, C. T., & Miller, D. J. (2013). A novel process for recovery of fermentation-derived succinic acid: Process design and economic analysis. *Bioresource Technology, 139*, 235–241.

Padma, P. N., Anuradha, K., Nagaraju, B., Kumar, V. S., & Reddy, G. (2012). Use of pectin rich fruit wastes for polygalacturonase production by *Aspergillus awamori* MTCC 9166 in solid state fermentation. *Journal of Bioprocessing and Biotechniques, 2*, 2.

Panouillé, M., Ralet, M. C., Bonnin, E., & Thibault, J. F. (2007). Recovery and reuse of trimmings and pulps from fruit and vegetable processing. In K. W. Waldron (Ed.), *Handbook of waste management and co-product recovery in food processing* (pp. 417–447). Cambridge: Woodhead Publishing Limited.

Pateraki, C., Patsalou, M., Vlysidis, A., Kopsahelis, N., Webb, C., Koutinas, A. A., et al. (2016). *Actinobacillus succinogenes*: Advances on succinic acid production and prospects for development of integrated biorefineries. *Biochemical Engineering Journal, 112*(2016), 285–303.

Pazmiño-Durán, E. A., Giusti, M. M., Wrolstad, R. E., & Glória, M. B. A. (2001). Anthocyanins from banana bracts (*Musa × paradisiaca*) as potential food colorants. *Food Chemistry, 73*, 327–332.

Pinazo, J. M., Domine, M. E., Parvulescu, V., & Petru, F. (2015). Sustainability metrics for succinic acid production: A comparison between biomass-based and petrochemical routes. *Catalysis Today, 239,* 17–24.

PRé Consultants. (2017). *SimaPro database manual-methods library.* The Netherlands.

Ramesh, T., & Kalaiselvam, M. (2011). An experimental study on citric acid production by *Aspergillus niger* using Gelidiella Acerosa as a substrate. *Indian Journal of Microbiology, 51,* 289–293.

Ribeiro da Silva, L. M., Teixeira de Figueiredo, E. A., Silva Ricardo, N. M., Pinto Vieira, I. G., Wilane de Figueiredo, R., Brasil, I. M., et al. (2014). Quantification of bioactive compounds in pulps and by-products of tropical fruits from Brazil. *Food Chemistry, 143,* 398–404.

Ruiz, E., Gullón, B., Moura, P., Carvalheiro, F., Eibes, G., Cara, C., et al. (2017). Bifidobacterial growth stimulation by oligosaccharides generated from olive tree pruning biomass. *Carbohydrate Polymers, 169,* 149–156.

Sabeena Farvin, K. H., Grejsen, H. D., & Jacobsen, C. (2012). Potato peel extract as a natural antioxidant in chilled storage of minced horse mackerel (*Trachurus trachurus*): Effect on lipid and protein oxidation. *Food Chemistry, 131,* 843–851.

Sagar, N. A., Pareek, S., Sharma, S., Yahia, E. M., & Lobo, M. G. (2017). Fruit and vegetable waste: Bioactive compounds, their extraction, and possible utilization. *Comprehensive Reviews in Food Science and Food Safety, 17,* 512–531.

Sala, S., Anton, A., McLaren, S. J., Notarnicola, B., Saouter, E., & Sonesson, U. (2017). In quest of reducing the environmental impacts of food production and consumption. *Journal of Cleaner Production, 140,* 387–398.

Sandhya, R., & Kurup, G. (2013). Screening and isolation of pectinase from fruit and vegetable wastes and the use of orange waste as a substrate for pectinase production. *International Research Journal of Biological Sciences, 2,* 34–39.

Sauer, M., Porro, D., Mattanovich, D., & Branduardi, P. (2008). Microbial production of organic acids: expanding the markets. *Trends in Biotechnology, 26,* 100–108.

Searcy, C. (2012). Corporate sustainability performance measurement systems: A review and research agenda. *Journal of Business Ethics, 107,* 239–253.

Seifi, M., Seifi, P., Hadizadeh, F., & Mohajeri, S. A. (2013). Extraction of lycopene from tomato paste by ursodeoxycholic acid using the selective inclusion complex method. *Journal of Food Science, 78,* 1680–1685.

Singh, A., Sabally, K., Kubow, S., Donnelly, D. J., Gariepy, Y., Orsat, V., et al. (2011). Microwave-assisted extraction of phenolic antioxidants from potato peels. *Molecules, 16,* 2218–2232.

Smidt, M., den Hollander, J., Bosch, H., Xiang, Y., van der Graaf, M., Lambin, A., et al. (2016). Life cycle assessment of biobased and fossil-based succinic acid. In J. Dewulf, S. De Meester & R. Alvarenga (Eds.), *Sustainability assessment of renewables-based products: Methods and case studies* (pp. 307–322).

Spangenberg, J. H. (2004). Reconciling sustainability and growth: Criteria, indicators, policies. *Sustainable Development, 12,* 74–86.

Teixeira, A., Baenas, N., Dominguez-Perles, R., Barros, A., Rosa, E., Moreno, D. A., et al. (2014). Natural bioactive compounds from winery by-products as health promoters: A review. *International Journal of Molecular Sciences, 15*(9), 15638–15678.

UNEP-SETAC. (2009). In C. Benoît Norris & B. Mazijn (Eds.), *Guidelines for social life cycle assessment of products.* Paris, France: United Nations Environment Programme (UNEP). Available from: http://www.unep.org/publications/.

United Nations. (2015). *Time for global action for people and planet. Sustainable development knowledge platform.* United Nations Department of Economic and Social Affairs. Available from: https://sustainabledevelopment.un.org/post2015.

Van Maris, A. J., Konings, W. N., van Dijken, J. P., & Pronk, J. T. (2004). Microbial export of lactic and 3-hydroxypropanoic acid: Implications for industrial fermentation processes. *Metabolic Engineering, 6,* 245–255.

Vilas-Boas, S. G., Esposito, E., & Mendonca, M. M. (2002). Novel lignocellulolytic ability of *Candida utilis* during solid-substrate cultivation on apple pomace. *World Journal of Microbiology & Biotechnology, 18,* 541–545.

Wang, X., Chen, Q., & Lü, X. (2014). Pectin extracted from apple pomace and citrus peel by subcritical water. *Food Hydrocolloids, 38,* 129–137.

Weidema, B. P. (2006). The integration of economic and social aspects in life cycle assessment. *International Journal of Life Cycle Assessment, 11,* 89–96.

Wernet, G., Bauer, C., Steubing, B., Reinhard, J., Moreno-Ruiz, E., & Weidema, B. (2016). The ecoinvent database version 3 (part I): overview and methodology. *The International Journal of Life Cycle Assessment, 21,* 1218–1230.

Werpy, T., & Petersen, G. (2004). *Top value added chemicals from biomass—Results of screening for potential candidates from sugars and synthesis gas,* vol. I, U.S.Department of Energy, Oak Ridge, http://www.osti.gov/bridge.

Wiloso, E. I., Heijungs, R., & Huppes, G. (2014). A novel life cycle impact assessment method on biomass residue harvesting reckoning with loss of biomass productivity. *Journal of Cleaner Production, 81,* 137–145.

Zhang, Q., Cheng, C. L., Nagarajan, D., Chang, J. S., Hu, J., & Lee, D. J. (2017). Carbon capture and utilization of fermentation CO_2: Integrated ethanol fermentation and succinic acid production as an efficient platform. *Applied Energy, 206,* 364–371.

Zheng, Z., & Shetty, K. (2000). Enhancement of pea (*Pisum sativum*) seedling vigour and associated phenolic content by extracts of apple pomace fermented with *Trichoderma* spp. *Process Biochemistry, 36,* 79–84.

Dr. Sara González-García Chemical engineer as degree (USC, 2005), she got her Ph.D. in 2009. She is postdoctoral researcher at the University of Santiago de Compostela. Her main research topics include the sustainability assessment applying life-cycle assessment and carbon footprint in industrial, forestry, agricultural and food sectors. These research activities have produced two Ph. D. theses (currently, she is co-director of three Ph.D. theses), 90 papers published in international journals, 7 book chapters and 3 national publications. She has an h-index of 26 (with 1860 total citations by SCOPUS).

Dr. Patricia Gullón Food Science and Technology as degree (UVigo, 2006), she got her Ph.D. in 2011. She is postdoctoral researcher at the University of Basque Country. Her main research line is focused on the obtaining of value-added bioproducts from lignocellulosic materials under an integral biorefinery perspective. She is co-director of two Ph.D. theses (ongoing), two master theses (completed). Her CV includes 32 papers published in international journals, 6 book chapters, 3 proceedings as well as more 50 contributions to international conferences. She has an h-index of 15 (with 946 total citations by SCOPUS).

Dr. Beatriz Gullón Food Science and Technology as degree (UVigo, 2006), she got her Ph.D. in 2011. She is postdoctoral researcher at the University of Santiago de Compostela. Her main research topic is focused on the exploitation of natural compounds, by-products and derived extracts in terms of their potential biological properties. Her CV includes 59 papers published in international journals, 6 book chapters as well as more 65 contributions to international conferences. She has an h-index of 18 (with 726 total citations by SCOPUS).

Environmental Indicators in the Meat Chain

Ilija Djekic and Igor Tomasevic

Abstract This chapter gives an overview of the main environmental indicators in the meat chain. The meat sector is considered as one of the leading polluters in the food industry where its impact affects the entire meat chain. Regardless of the research methodology, environmental impacts of the meat chain occurs in three dimensions—climate change, revealing the necessity of analyzing greenhouse gas emissions in perspectives of global warming potential, consumption of natural resources mainly water and energy, and polluting the environment with waste (both organic and inorganic) and polluted wastewater. Bottom-up approach in analyzing environmental indicators provides new evidence relating to the meat sector. It can help environmental specialists and managers in the meat sector, directing them as to how to improve environmental practices on-site. Finally, this chapter gives an overview of improvement perspectives and future research dimensions.

Keywords Meat chain · Environmental indicators · Life cycle
Environmental impact · Environmental footprints

1 Introduction

Human beings have a long history of consuming meat, and meat products are considered as omnivores. The first human beings were scavengers and/or hunters (Speth 1989). Depending on the type of animal, carnivores (meat eaters) have digestive systems equipped to fully consume and use animal foods whether through predation or scavenging. On the other side, herbivores (plant eaters) have

I. Djekic (✉)
Department of Food Safety and Quality Management, Faculty of Agriculture,
University of Belgrade, Nemanjina 6, 11080 Belgrade, Republic of Serbia
e-mail: idjekic@agrif.bg.ac.rs

I. Tomasevic
Department of Animal Origin Products Technology, Faculty of Agriculture,
University of Belgrade, Nemanjina 6, 11080 Belgrade, Republic of Serbia

© Springer Nature Singapore Pte Ltd. 2019
S. S. Muthu (ed.), *Quantification of Sustainability Indicators in the Food Sector*,
Environmental Footprints and Eco-design of Products and Processes,
https://doi.org/10.1007/978-981-13-2408-6_3

specialized organs to digest cellulose such as bovines (sheep, deer, goats, etc.), equines (horses), and lagomorpha (rabbits and hares). Consequently, eating meat from herbivores is an efficient way for humans to indirectly make the most of plants, grass, and any type of natural pasture. It has been recorded that human ancestors were eating meat as early as 1.5 million years ago (Domínguez-Rodrigo et al. 2012). Since then, humanity is consuming meat from different types of animals and meat consumption became part of our culture.

As a result of world's population growth and overall consumption of meat per capita, it is obvious that meat production is increasing every year (Henchion et al. 2014). One of the reasons for expanding meat production is trade liberalization and globalization of food systems (Delgado 2003). The second reason may be found within the nutritional needs and accepted dietary patterns by consuming foods with higher content in animal protein (Hawkesworth et al. 2010; Mathijs 2015). Finally, consumers worldwide are fond of meat products mainly because of their sensory attributes and cultural habits (Font-i-Furnols and Guerrero 2014).

On the other side, meat is considered as a type of food product holding the greatest environmental impact throughout the food chain (Röös et al. 2013). Regardless of the perspective, environmental impacts of this chain influences three dimensions: (i) climate change in respect to the global warming potential, acidification potential and eutrophication potential; (ii) consumption of natural resources mainly water and energy; and (iii) polluting the environment with discharge of wastewater and various types of waste (Djekic 2015). In order to compare the environmental performances over time and against other companies in the meat chain, it is necessary to develop and make available environmental indicators (Đekić and Tomašević 2017b; Jasch 2000). Due to the economic, environmental, and social implications of the meat chain, meat production, and meat consumption are linked to the three sustainability pillars—economy, society, and environment (Allievi et al. 2015).

Mapping the process(es) and setting the scope and boundaries are important in order to clarify environmental impacts of the food chain analyzed from a "farm to fork" perspective (Djekic et al. 2018). Wider perspective of the meat chain identifies five main stakeholders: (i) farm(er)s, (ii) slaughterhouses, (iii) meat processors, (iv) customers (HoReCa, supermarkets, butcheries, retailers), and (v) consumers (Borrisser-Pairó et al. 2016; Djekic and Tomasevic 2016).

"Farming" is the first stage in the meat chain and it covers all livestock activates where the two major environmental contributors are feed production and waste/manure management (McAuliffe et al. 2016). "Slaughterhouse" covers reception of live animals, livestock handling, animal welfare, slaughtering, and chilling while "meat processing plant" start at the incoming control of carcasses and ends up with the storage of (processed) meat products, including but not limited to thermal meat processing and waste handling (Djekic et al. 2015). Under certain occasions, these two stages can take place at the same premises. Main environmental impacts in slaughterhouses and meat processing plants are usage of energy, usage of water, waste handling, and wastewater discharge (Djekic et al. 2016). "Customers" are recognized as points of sale of meat and meat products such as supermarkets, grocery shops, or butcher's shops (Djekic and Tomasevic 2016). Finally, "consumers" are considered as the final link the meat chain and household use covers all activates after purchasing of meat and meat products such as refrigeration of meat (Coulomb 2008), meat preparation, and cooking (Xu et al. 2015), as well as discharge of packaging waste and bio-waste (Skunca et al. 2018).

The objective of this chapter is to give an overview of the main environmental indicators in the meat chain. Section 2 gives an overview of the meat production in the world. Section 3 analyses environmental indicators that exist in the meat chain, deployed in three levels of indicators. Section 4 shows generic meat chain indicators and further deploys them from a case study perspective for pork meat, beef meat, and poultry meat. Section 5 analyses environmental impacts of the meat chain highlighting routes to improvements. Concluding remarks are given in Sect. 6.

2 Meat Production

Overall world meat production is estimated at around 320 million tons in 2016, with a growth in the Americas and Europe and a slight downturn in China and Australia (OECD/FAO 2017b). Among various meat sectors, poultry, and bovine meat production expanded, while pig meat and sheep meat production have expressed a slight decline. The poultry sector expanded, coming in at more than 117 million tons in 2016 with a forecast of nearly 118 million tons in 2017 (Table 1).

Excluding China, aggregate meat production of the rest of the world is expected to rise by almost 2.0% year on year. Deployed by category, bovine meat is expected to show the largest growth in production, with marginal increases for poultry and ovine meat, and a slight fall for pork meat. The global meat trade has recovered during the year 2016, rising by 5% to 30 million tonnes.

Table 1 World balance for meats by type (OECD/FAO 2017b)

	2015	2016 (estimate)	2017 (forecast)	Change: 2017 over 2016
	Million tonnes			%
Production	**320.5**	**321.0**	**322.0**	**0.3**
Bovine meat	67.6	68.3	69.6	1.6
Poultry meat	116.9	117.2	117.7	0.4
Pig meat	116.1	115.6	114.7	−0.8
Ovine meat	14.4	14.4	14.5	0.6
Trade	**29.9**	**31.2**	**32.0**	**2.5**
Bovine meat	9.2	8.9	9.0	0.8
Poultry meat	12.2	12.8	13.2	2.9
Pig meat	7.2	8.3	8.6	4.1
Ovine meat	1.0	0.9	0.9	−2.0

Numbers in bold present overall share of production and trade of meat

Global meat production is projected to be 13% higher in 2026 compared to the base period (2014–2016). Developing countries will mainly influence the total increase and consequently will have a more intensive use of feed in the production process. Poultry meat is recognized as the primary driver of the growth in total meat production, in response to expanding global demand for this more affordable animal protein compared to red meats. Low production costs and lower product prices are the main triggers to making poultry becoming the most favorable meat for both producers and consumers in developing countries (OECD/FAO 2017a).

Over the last 50 years, global meat consumption rose from 23.1 kg per person per year in 1961 to 42.2 kg per person per year in 2011 (Sans and Combris 2015). Meat consumption worldwide per capita is expected to stagnate at 34.6 kg retail weight equivalent by 2026. In relation to the population growth rates in the developing world, total consumption is expected to increase by nearly 1.5% per annum (OECD/FAO 2017a).

Driven by economic development and urbanization over the last 50 years, animal-based protein consumption has increased worldwide, rising from 61 g per person per day in 1961 to 80 g per person per day in 2011 (Sans and Combris 2015). It is estimated that 1–9% of human beings are vegetarians in developed countries and 40% in India (Ruby 2012). Flexitarians (person who eats mostly as a vegetarian but sometimes includes meat, fish, or poultry) are more and more numerous. They have different moral drivers than vegetarians raising concerns about animal welfare more than full-time meat eaters but less than vegetarians (De Backer and Hudders 2015). Taking into account the vegetarians and other voluntary dietary habits such as veganism (exclusion from animal products), raw foodism (dietary practice of eating only uncooked, unprocessed foods), fruitarianism (diet that consists entirely or primarily of fruits and possibly nuts and seeds, without any animal products), and various religious restrictions, we can say that a large majority of the human population eat meat regularly or occasionally. In summary, humanity used to and still relies on meat and meat products.

3 Environmental Indicators in the Meat Chain

Evaluation of environmental impacts depends on the approach and methods used (Carvalho et al. 2014). The very common approach is by introducing and calculating environmental performance indicators (EPIs). EPI is "a measurable representation of the status of operations, management or conditions related to environmental aspects" (ISO 2015). Henri and Journeault (2008) highlight two main reasons for calculating reliable numeric indicators: organization's legal responsibility on environmental issues and achievement of certain environmental objectives. They also conclude that financial indicators are understood as backward-looking, with limited ability to explain environmental performance.

Fig. 1 Levels of EPIs

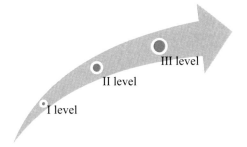

Rule of the thumb for all EPIs are that they should be (i) measurable; (ii) objective; (iii) verifiable; (iv) repeatable; and (v) technically feasible (Đekić 2009). In general, there are three levels of EPIs that are related to the maturity of implemented environmental practice (Đekić and Tomašević 2017a) Fig. 1.

3.1 First-Level Indicators

The first level of EPIs is basic indicators with numerical values (Đekić and Tomašević 2017a). They can be divided into two categories with no connection between the indicators. The first category consists of generic environmental indicators such as energy and water consumption, wastewater discharge. The second category comprises of indicators related to meat production. Some first-level EPIs related to meat production are presented in Table 2.

3.2 Second-Level Indicators

The second-level EPIs are calculated from at least two first-level EPIs. In order to evaluate food production, it is necessary to define a unit in which the impacts are

Table 2 First-level EPIs

Generic environmental EPIs		Meat production EPIs	
Indicator	Unit	Indicator	Unit
Consumption of electric energy	MJ/kWh	Livestock production	t/kg
Consumption of thermal energy	MJ/kWh	Carcass production	t/kg
Consumption of fossil fuels	t/m^3	Fresh meat production	t/kg
Consumption of water	m^3/L	Production of meat products	t/kg
Consumption of chemicals	L/kg	Consumption of additives	t/kg
Wastewater discharge	m^3/L	Consumption of spices	t/kg
Waste discharge	t/kg	Consumption of packaging	units/kg

presented such as 1 kg of food, and to define a formula for calculation of this type of EPIs. The functional unit (FU) is the unit to which the results are expressed and a basis for comparisons (Djekic et al. 2018). This approach to environmental performance shows the relationship between production performance and the environment, depicting environmental impacts of the processes within a meat company (Dubey et al. 2015).

Proper choice of the FU is of utmost importance since different functional units can lead to different results for the same (meat) production systems (Djekic and Tomasevic 2016). In the meat chain, the most common FUs are 1 kg of livestock (Basset-Mens and van der Werf 2005; Dalgaard et al. 2007); 1 kg of carcass (Nguyen et al. 2011; Williams et al. 2006); and 1 kg of meat/meat products (Cederberg and Flysjö 2004).

Most common second level of EPIs in the meat production are meat yield (share of lean meat in live animals and/or in carcass), solid output per FU (in farming mostly manure, in slaughtering/deboning percentage of by-product such as offal, bones, fat, and skin), and resource consumption per FU (energy/water) (Đekić and Tomašević 2017a). Focus in the meat chain is calculation of consumptions and discharges per meat FUs such as energy-to-meat ratio, water consumption per meat product, wastewater discharge per meat product, and chemical usage per FU (Djekic et al. 2015; IPPC 2006; UNEP 2000). Table 3 gives an overview of the most common second-level EPIs in the meat chain.

Further deployment of this indicator can help in specifying environmental impacts within the meat chain. Energy is used in all parts of the meat chain basically for machines and equipment, for controlling temperature regimes (heating/ refrigerating), and for transportation purposes (Djekic 2015; IPPC 2006). Energy deployment should go towards clarifying consumption of electric energy, thermal energy, and other sources of energy such as types and quantities of fossil fuels. Water is very important throughout the meat chain. It is necessary for live animals at farms and when entering the slaughterhouse and plays a significant role in hygiene and sanitation of slaughterhouses, meat processing plants, and retail. Finally, it is used at households for meat preparation (Djekic and Tomasevic 2016; IPPC 2006).

Waste discharge can determine types and quantities of waste (organic vs. inorganic waste, hazardous vs. nonhazardous waste, etc.). Types of waste can further be separated depending on the waste material (plastic, metal, wooden, paper, food waste, and cardboard). Within the meat chain, there are two main types of wastes—inedible products (bones, fat, heads, legs, skins, hair, and offal) and various packaging materials (Djekic et al. 2016; Kupusovic et al. 2007). Some quantities of organic waste are a result of consumer demands. They prefer lean meat, which causes the production of (organic) waste in both slaughterhouses and meat processing plants (Rahman et al. 2014). Handling this type of animal by-products is regulated by the law in developed markets, like in the EU (EC 2009).

Wastewater is a result of cleaning and sanitation and covers washing of livestock, carcasses, and offal, cleaning and sanitation of equipment and work environment and workers' personal hygiene (Kupusovic et al. 2007). At slaughterhouses, when

Table 3 Overview of the most common second-level EPIs

Meat chain	Indicator	Formula (unit)
Farm	Consumption of water per FU	$\dfrac{\text{Consumption of water (L)}}{\text{FU (kg of livestock)}}$
	Consumption of energy per FU	$\dfrac{\text{Consumption of energy (MJ)}}{\text{FU (kg of livestock)}}$
	Consumption of fossil fuels per FU	$\dfrac{\text{Consumption of fuels (L)}}{\text{FU (kg of livestock)}}$
	Discharge of wastewater per FU	$\dfrac{\text{Discharge of wastewater (L)}}{\text{FU (kg of livestock)}}$
	Discharge of waste per FU	$\dfrac{\text{Discharge of waste (kg)}}{\text{FU (kg of livestock)}}$
Slaughter house	Consumption of water per FU	$\dfrac{\text{Consumption of water (L)}}{\text{FU (kg of carcass)}}$
	Consumption of energy per FU	$\dfrac{\text{Consumption of energy (MJ)}}{\text{FU (kg of carcass)}}$
	Consumption of fossil fuels per FU	$\dfrac{\text{Consumption of fuels (L)}}{\text{FU (kg of carcass)}}$
	Discharge of wastewater per FU	$\dfrac{\text{Discharge of wastewater (L)}}{\text{FU (kg of carcass)}}$
	Discharge of waste per FU	$\dfrac{\text{Discharge of waste (kg)}}{\text{FU (kg of carcass)}}$
Meat processing	Consumption of water per FU	$\dfrac{\text{Consumption of water (L)}}{\text{FU (kg of meat product)}}$
	Consumption of energy per FU	$\dfrac{\text{Consumption of energy (MJ)}}{\text{FU (kg of meat product)}}$
	Consumption of fossil fuels per FU	$\dfrac{\text{Consumption of fuels (L)}}{\text{FU (kg of meat product)}}$
	Discharge of wastewater per FU	$\dfrac{\text{Discharge of wastewater (L)}}{\text{FU (kg of meat product)}}$
	Discharge of waste per FU	$\dfrac{\text{Discharge of waste (kg)}}{\text{FU (kg of meat product)}}$
Retail	Consumption of water per FU	$\dfrac{\text{Consumption of water (L)}}{\text{FU (kg of meat product)}}$
	Consumption of energy per FU	$\dfrac{\text{Consumption of energy (MJ)}}{\text{FU (kg of meat product)}}$
	Discharge of waste per FU	$\dfrac{\text{Discharge of waste (kg)}}{\text{FU (kg of meat product)}}$
Household	Consumption of water per FU	$\dfrac{\text{Consumption of water (L)}}{\text{FU (kg of meat product)}}$
	Consumption of energy per FU	$\dfrac{\text{Consumption of energy (MJ)}}{\text{FU (kg of meat product)}}$
	Discharge of waste per FU	$\dfrac{\text{Discharge of waste (kg)}}{\text{FU (kg of meat product)}}$

FU functional unit. In meat industry it is 1 kg of livestock or 1 kg of carcass or 1 kg of meat product (depending on the role in the meat chain) discharged, water is an effluent with high organic loads coming from manure, blood, and fat and undigested stomach contents (UNEP 2000). Beyond quantity of wastewater, this indicator can analyze wastewater load and quality of wastewater such as values for biological oxygen demand and/or chemical oxygen demand in terms of FU.

3.3 Third-Level Indicators

The third level of EPIs provides information on different environmental footprints (Đekić and Tomašević 2017a). Environmental footprint is a quantitative measurement that calculates or describes the misuses of natural resources by humans (Hoekstra 2008). Footprint tools are tools for footprint calculations and suggested reduction in terms of prevention of pollution or environmental improvement (Čuček et al. 2015).

Three most recognized members of the footprint family are ecological, water, and carbon footprints (Herva et al. 2011). The ecological footprint is related to the natural, social, cultural, and economic environment and is not commonly calculated in the meat chain (Đekić and Tomašević 2017a). It refers to the number of individuals who can be supported in a given area within natural resource limits, without degrading the environment for present and future generations (Kratena 2008). The water footprint is built on the concept of virtual water related to all links in the meat chain and refers to total water used during the production of all goods and services in the entire meat chain (Herva et al. 2011). It consists of blue (consumption of surface and groundwater), green (consumption of rainwater stored within the soil as soil moisture), and gray (volume of freshwater required for assimilating the load of pollutants based on existing ambient water quality standards) water footprints (Čuček et al. 2015; Mekonnen and Hoekstra 2010). As presented before, meat companies calculate various second-level EPIs related to water consumption so further calculations of this footprint are possible.

Carbon footprint measures all greenhouse gas emissions caused directly and indirectly and is expressed in CO_2 equivalent since the largest single contributor to climate change is CO_2 (Herva et al. 2011). The predominant greenhouse gases (GHG) emitted from agriculture are methane (CH_4) and nitrous oxide (N_2O). They possess 21 and 310 times of the global warming potential (GWP) of carbon dioxide, respectively (IPPC 2006; MacLeod et al. 2013). The main GHGs are CO_2, CH_4, N_2O, hydrochlorofluorocarbons (HCFCs), hydrofluorocarbons (HFCs), and ozone in the lower atmosphere (WMO 2017). It is considered that emission of GHG leads to increased droughts, floods, losses of polar ice caps, sea-level rising, soil moisture losses, forest losses, changes in wind and ocean patterns, and changes in agricultural production (Čuček et al. 2015).

Influence of climatic conditions on food safety, incidence, and prevalence of food-borne diseases becomes an important connection between climate change and the food chain (Bezirtzoglou et al. 2011; Holvoet et al. 2014; Lal et al. 2012;

Miraglia et al. 2009). Temperature and precipitation changes and patterns, both locally and globally, are related with the transport, growth, and survival of enteric bacteria (Liu et al. 2015). Most of the published publications related to climatic condition and food safety, are from the farms (Holvoet et al. 2014; Kirezieva et al. 2015; Liu et al. 2013; Uyttendaele et al. 2015). Also, intensive precipitations are linked with contamination pathway of pathogens in the meat chain such as from manure at livestock farms and from grazing pastures (Parker et al. 2010) as well as the microbial contamination of vegetables coming from fecal waste into the soil or contaminated water (Holvoet et al. 2014).

It is important to note that the majority of environmental footprints and models used to evaluate environmental impacts were developed by environmental scientists. They all are generic regardless of the type of companies or products with limited environmental models/footprints for the food industry and with no specific model/footprint tailored for the meat chain (Djekic et al. 2018).

Latest researches confirm that carbon footprint is used in presenting environmental impact of the meat chain (Đekić and Tomašević 2017a). Livestock and activities at the farms contribute to global warming potential directly coming from enteric fermentation and manure management and indirectly as a result of feed production (Gerber et al. 2015; Röös et al. 2013). The global warming potential within the meat chain can be calculated as follows:

$$\text{GWP} = \sum_{i}^{n} \text{GWP}_i x m_i \left[\text{kgCO}_{2eq} \right]$$

where m_i—mass of emitted gas (kg) and GWP_i—global warming potential of the emitted gas. The GWP is usually calculated for every part of the meat chain.

Acidification potential is an indicator that calculates the potential of acidifying pollutants (SO_2, NO_x, HCl, NH_3, and HF) to form H^+ ions and damage plants, animals, and the ecosystem (Čuček et al. 2015). Ammonia is the main source of acidifying emissions during animal production released from manure in farms and during manure handling (Djekic et al. 2015). Liquid manure handling systems emit less ammonia than solid manure but liquid/slurry storage stimulates CH_4 production, due to anaerobe conditions (IPCC 2006). This potential is usually expressed in SO_2 equivalents. The acidification potential within the meat chain can be calculated as follows:

$$\text{AP} = \sum_{i}^{n} \text{AP}_i x m_i [\text{kg SO}_{2e}]$$

where m_i—mass of emitted substance (kg) and AP_i—acidification potential of the emitted substance. The AP is usually calculated for every part of the meat chain.

Eutrophication potential increases the aquatic plant growth attributable of nutrients left by overfertilization of water and soil such as nitrogen and phosphorus (Čuček et al. 2015). At the farm level, nitrates are accumulated during feed production and ammonia release from manure handling and as such dominate the emissions of eutrophying substances. It is considered as the main contributors to eutrophication within the meat chain (Röös et al. 2013). This potential is expressed in PO_3^{4-} equivalents. The eutrophication potential within the meat chain can be calculated as follows:

$$EP = \sum_i^n EP_i x m_i [kg\, PO_{4e}]$$

where m_i—mass of emitted substance (kg) and EP_i—eutrophication potential of the emitted substance. The EP is usually calculated for every part of the meat chain.

Ozone depletion potential is expressed as CFC-11 or R11 equivalents and is calculated as the potential for reducing the protective stratospheric ozone layer where ozone-depleting substances are freons, chlorofluorocarbons, carbon tetrachloride, and methyl chloroform (Čuček et al. 2015). It is known that keeping products at low temperatures inhibits the growth of potentially harmful microorganisms (Sofos 2014) and that the cold chain plays a significant role in keeping meat safe. The effectiveness of the cold chain depends on the time/temperature ratio, the kind of refrigerators, and the position of meat/meat products within it (Baldera Zubeldia et al. 2016). However, the cold chain requirements have an impact on ozone layer depletion due to the use of refrigerants for chilling/freezing and affect the entire meat chain (Djekic et al. 2015). The ozone depletion potential within the meat chain can be calculated as follows:

$$ODP = \sum_i^n ODP_i x m_i [kg\, R11]$$

where m_i—mass of emitted gas (kg) and ODP_i—ozone depletion potential of the emitted gas. The ODP is usually calculated for every part of the meat chain.

4 Meat Chain Indicators—Case Study

Life-cycle assessment (LCA) is considered as the best method in calculating environmental impact from all stages of agricultural and food production (Djekic 2015). The methodology is outlined in an international standard (ISO 14040:2006) and comprises of the following steps: (i) mapping the process, (ii) setting scope and boundaries, (iii) collecting inventory data, and (iv) interpreting the results (ISO 2006). Mapping the process joint with setting the scope and boundaries is to clarify which part of the meat chain is analyzed from the "farm to the fork" perspective (Djekic 2015). Collecting inventory data is the most important but the most challenging part, since uncertainty may occur due to imprecise data. Analysis of inventory requires calculation of environmental impacts defined in the goal of the

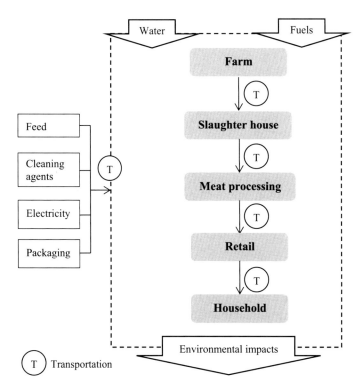

Fig. 2 Generic system boundaries of the meat chain. Gray boxes are premises where the environmental impacts occur

LCA in order to determine potential environmental impacts (McAuliffe et al. 2016). Interpretation of the results is the final stage that enables mitigation strategies in relation to environmental improvements. Generic system boundaries of the meat chain are presented in Fig. 2.

In order to convert data from the "whole of subsystem basis" to a "functional unit basis", it is necessary to allocate inputs and outputs. For this purpose, it is common to use one of the three main allocation methods economic allocation, physical allocation, and system expansions (de Vries and de Boer 2010). Still, there are differences in LCA model assumptions, system boundaries taken into account, functional units defined within the meat chain, data collection methods and data processing, environmental impact categories and emission factors, normalization methods, and weighting factors which make comparisons difficult (Carvalho et al. 2014; Djekic and Tomasevic 2016; Pennington et al. 2004).

Table 4 gives a generic overview of inventory data needed to perform a LCA of the entire meat chain.

Due to the fact that the highest level of environmental impacts occurs at the farms, some other important issues that should be considered, and wherever possible included in the LCA, are:

Table 4 Global inventory for the production of meat (entire meat chain)

	Unit	Subsystem				
		1	2	3	4	5
Input — materials						
Water	L/kg FU	●	●	●	●	●
Cleaning agents (alkaline)	L or g/kg FU		●	●	○	
Cleaning agents (acid)	L or g/kg FU		●	●	○	
Feed	kg/kg FU	●				
Input — energy						
Electric energy	kWh or MJ/kg FU	●	●	●	●	●
Thermal energy	MJ/kg FU		○	○	○	
Fossil fuels						
LPG	kg/FU	●	●	●	●	○
Natural gas	m^3/FU	●	●	●	●	○
Petrol	L/FU	●	●	●	●	○
Diesel	L/FU	●	●	●	●	○
Packaging materials						
HDPE	g/kg FU			●	●	○
PVC	g/kg FU			●	●	○
PET	g/kg FU			●	●	○
PP	g/kg FU			●	●	○
PE	g/kg FU			●	●	○
Cardboard/paper	g/kg FU			●	●	○
Aluminium foil	g/kg FU			●	●	○
Shrink foil	g/kg FU			●	●	○
Styrofoam	g/kg FU			●	●	○
Refrigerants	g/kg FU		●	●	●	○
Cooking oil	mL/kg FU					●
Output						
Production						
Annual production of meat	kg or tonnes	●	●	●		
Annual sale of meat	kg or tonnes				●	
Annual consumption of meat	kg					●

	Unit	Subsystem				
		1	2	3	4	5
Waste						
Waste —confiscate (food waste)	kg/kg FU	●	●	●	●	●
Waste — manure / slurry	m³/kg FU	●	●			
Other types of waste						
HDPE	g/kg FU			●	●	●
PVC	g/kg FU			●	●	●
PET	g/kg FU			●	●	●
PP	g/kg FU			●	●	●
PE	g/kg FU			●	●	●
Cardboard / paper	g/kg FU			●	●	●
Aluminium foil	g/kg FU			●	●	●
Shrink foil	g/kg FU			●	●	●
Wood	g/kg FU		●	●	●	
Waste water	L/kg FU	●	●	●	○	○

Subsystem 1 farm; *subsystem 2* slaughterhouse; *subsystem 3* meat processing plant; *subsystem 4* retail; *subsystem 5* household
FU functional unit: 1 kg of livestock or 1 kg of carcass or 1 kg of meat product (depending on the role in the meat chain)
HDPE high density polyethylene; *PVC* polyvinyl chloride; *PET* polyethylene terephthalate; *PP* polypropylene; *PE* Polyethylene
Signs: ● mandatory data; ○ data "nice to have"

- Type of breed.
- Type of production system.
- Pre-fattening period.
- Slaughtering age and weight.
- Male–female ratio.
- Mortality rate.
- Replacement rate.
- Feed production.
- Feed formulation.
- Direct on-farm emissions (N_2O, CH_4, NH_3, NO_3, PO_4, etc.).
- Good agricultural/veterinary practice in place.

At slaughterhouses, some of the issues that should also be considered are:

– Wastewater treatment system in place.
– Quality of wastewater (biochemical oxygen demand, chemical oxygen demand, total organic carbon, total suspended solids, total nitrogen, and phosphorus, etc.).
– Waste management in place.
– Maintenance of equipment and infrastructure.
– Good manufacturing/hygiene practice.

Meat processing plants should take into account:

– Maintenance of equipment and infrastructure.
– Meat product portfolio.
– Allocation factors of inventory to each type of meat product.
– Good manufacturing/hygiene practice.

At retail, issues to be considered are:

– Size of retail.
– Type of products sold at retail.
– Allocation factors of inventory to each type of meat product.
– Good retail/hygiene practice.

At households, the following should be investigated:

– Purchasing habits.
– Consumption patterns at home.
– Food preparation habits.
– Dietary issues.
– Allocation factors of meal preparation related to meat/meat product.

Overall, the good environmental practice should be evaluated at each stage of the meat chain. This is important since the level of environmental practices in respect to the size of meat companies shows that smaller companies have a lower level of environmental practice in place. They usually take environmental actions only as a reaction to threats and sanctions from legal authorities (Djekic et al. 2016). The absence of any environmental practice is due to the lack of knowledge and experience and limited resources (Santos et al. 2011). Other criteria that affect environmental practices are the parts of the meat chain in which they operate and whether they have a certified environmental management system (Djekic et al. 2016).

Finally, in order to fully understand LCA as a methodology, Table 5 gives advantages and disadvantages of using LCA in meat the meat chain, modified from (Čuček et al. 2015; Djekic et al. 2018; Djekic and Tomasevic 2016).

Main advantages are that this methodology is accepted worldwide and many scientific papers justify this methodology where the number of papers in the food

Table 5 Advantages and disadvantages of using LCA in the meat chain

Advantages	Disadvantages
Accepted in science	Difficulty in collecting data
Standardized method (ISO 14040)	Uncertainty of collected data
Useful for the whole meat chain	Inventory influences results
Identifies "critical" spots	Allocation method influences results
Potential for eco-labelling	Limitation of study influences results
Optimization within a life cycle	System boundaries influence results
Technology comparison	Functional units influence results
	May be subjective
	Different software solutions

industry is increasing (Djekic et al. 2018). Standard ISO 14040 explains the methodology for performing an LCA study (ISO 2006). Due to the "cradle to grave" methodology outlined in LCA, it is very useful for analyzing food/meat chains. Good LCA can identify critical environmental spots that seek for optimization and environmental improvements. Calculations and results can be used in eco-labeling and marketing of meat product. Also, it can be used for technology comparisons.

On the other side, disadvantages are mainly related to the possibility to influence final results depending on the quality of data, allocation methods used, depth of inventory analysis, system boundaries used and functional units in which the results are presented. Since there is a large number of different software used, this may influence results and benchmarking of results. Finally, the focus of LCA is more on environmental impacts than on sustainability.

4.1 Pork Meat

The pig sector is considered as being one of the biggest contributors to global meat production, with over 37% and it is expected that global demand for pork meat will grow by over 35% until 2030 (MacLeod et al. 2013). Besides the economic part, the pork industry demands consumption of natural resources (water and energy) and generates remarkable waste flows (Noya et al. 2017). The evaluation of the contribution of pig production to environmental impacts is an ongoing story (Reckmann et al. 2012). Table 6 present a short summary of manuscripts covering LCA of least one part of the pork meat chain.

From the Table 6, it can be concluded that the majority of research was focused on farms, with a limited number of studies covering retail and households. The common potentials were global warming potential (GWP), acidification potential (AP), and eutrophication potential (EP) as well as energy consumption, mainly in retail.

Table 6 Summary of studies linking environmental impacts to pork meat chain

Authors	Research focus	System boundaries					Environmental impact
		1	2	3	4	5	
McAuliffe et al. (2016)	LCA of pig production	✓					GWP, AP, EP
de Vries and de Boer (2010)	LCA of livestock products	✓					GWP, AP, EP, LC, EC
Basset-Mens and van der Werf (2005)	LCA pig production	✓					EP, GWP, AP, EC, LC
Nguyen et al. (2012a)	LCA of pork production	✓					GWP, AP EP, HT, FEP, WC
Reckmann et al. (2012)	LCA of pork production	✓	✓				GWP, AP, EP, OLD, EC, LC
Djekic et al. (2015)	LCA of pork production	✓	✓	✓			GWP, EP, AP, OLD, HT
Roy et al. (2012)	LCA of meats (pork, beef, chicken)	✓	✓	✓	✓		GWP
Carlsson-Kanyama (1998)	LCA of food consumption	✓	✓	✓	✓	✓	GWP, EC

Subsystem 1 farm; *subsystem 2* slaughterhouse; *subsystem 3* meat processing plant; *subsystem 4* retail; *subsystem 5* household use

GWP global warming potential; *AP* acidification potential; *EP* eutrophication potential, *OLD* ozone layer depletion; *HT* human toxicity; *LC* land competition/use; *EC* energy consumption; *WC* water consumption; *FEP* fresh water aquatic ecotoxicity

FU: 1 kg of pork; 1 kg of bone and fat free meat; 1 kg of carcass; 1 kg of pork meat product

Results from review papers covering 20 pork meat-related LCA studies show the range of GWP per kg of bone-free meat (farms and slaughterhouses) from 3.6 to 8.9 kg CO_{2eq} (Cherubini et al. 2014; Röös et al. 2013). These papers emphasize differences in respect to countries (developed vs. developing), production systems (organic, conventional) as well as economic perspective (high and low profit). In Serbia, overall GWP throughout the life cycle (from farms to retails) is over 9 kg CO_{2eq} per kg of FU (Djekic et al. 2015) while European LCA studies show an average GWP of pork production of 3.6 kg CO_{2eq} per kg pork, ranging from 2.3 to 6.4 kg CO_{2eq} for different FU (1 kg of pork, 1 kg of bone, fat-free meat, and 1 kg of carcass) (Reckmann et al. 2012). Results from Japan show GHG emission of pork (farm gate, including manure) to be 5.57 CO_{2eq}/kg-meat while in slaughterhouses the GWP is estimated to be 0.12 kg CO_{2eq}/kg-meat (Roy et al. 2012).

Work from Röös et al. (2013) show that AP results are in the range from 0.026 to 0.156 (kg SO_{2eq}) covering farms and slaughterhouses with the FU being 1 kg of bone-free meat. On the other side, the analysis of nine pork production LCA studies performed by de Vries and de Boer (2010) show the range from 0.043 to 0.741 kg SO_2/kg.

EP results show a range from 0.015 to 0.102 kg PO_{4eq} (Röös et al. 2013). Similar to AP, de Vries and de Boer (2010) point that EP of the same product shows large variations.

As pointed above, the highest environmental impacts arise at the farm stage and latest research show segmentation of the pig production into piglet production and weaning to slaughtering (Basset-Mens and van der Werf 2005). The weaning to slaughtering stage contributes more since this stage lasts longer compared to piglet production and due to the weight of the pigs—the higher the weight, the more feed they eat and thus excrete more manure (Reckmann et al. 2012).

4.2 Beef Meat

Livestock production, particularly beef supply chain is considered as another major contributor to GHG emission of the meat chain (Bragaglio et al. 2018). In analyzing LCA of beef meat, many different circumstances have to be considered. The first issue is the production system. Some LCA cover analysis and/or comparison of intensively reared dairy calves and suckler herds (Nguyen et al. 2010). Other studies compare extensive cow-calf production, fattening system, cow-calf intensive, or traditional beef production system (Bragaglio et al. 2018). Finally, the introduction of "organic" production brings a new dimension (Buratti et al. 2017). Within the production systems, differences occur due to the origin of calves, duration of fattening period, diet formulation, etc. (Bragaglio et al. 2018). Other issues are related to the scope and system boundaries in terms of production of only beef meat, or production of beef meat and dairy products.

As Table 7 presents, the majority of studies are only focused on farms, specifically the production systems that are in place and comparison of the systems. Within the beef chain, common potentials were global warming and eutrophication as well as resource use (energy and water).

Due to the great variety in production systems, environmental impact of beef production showed the highest level of differences in results, when compared between each other. GHG emissions vary from 8.6 up to 35.2 kg CO_{2eq} per kg of edible beef while another interesting indicator, land use also varies from 12.1 to 47.2 m^2 (De Vries et al. 2015).

The belief that organic is always more environmental friendly was confirmed in works of Tsutsumi et al. (2018). GWP of organic production was 29.3 kg CO_{2eq}/kg of cold carcass steer weighs compared to 35.1 kg CO_{2eq}/kg of cold carcass steer weighs for conventional production. However, this was opposed by the research of Buratti et al. (2017). The organic system produces 24.6 kg CO_{2eq}/kg of live weight compared to the conventional that produces 18.2 kg CO_{2eq}/kg of live weight. The same study confirmed that enteric fermentation contributes with 50% of the total GHG emissions.

Table 7 Summary of studies linking environmental impacts to beef meat chain

Authors	Research focus	System boundaries					Environmental impact
		1	2	3	4	5	
Nguyen et al. (2010)	LCA beef production systems	✓					GWP, AP, EP, LC, EC
Bragaglio et al. (2018)	LCA beef production systems	✓					GWP, AP, EP, WC, LC
Buratti et al. (2017)	LCA beef production systems	✓					GWP
Tsutsumi et al. (2018)	LCA beef production systems	✓					GWP, AP, EP, EC
Ogino et al. (2016)	LCA beef production systems	✓					GWP, AP, EP, EC
Huerta et al. (2016)	LCA beef production	✓	✓		✓		GWP, AP, EP, LC, WC, RD, HT
Nguyen et al. (2012b)	LCA beef production systems	✓					GWP, AP, EP, EC
Dick et al. (2015)	LCA beef production systems	✓					GWP, LC, WC, RD

Subsystem 1 farm; *subsystem 2* slaughterhouse; *subsystem 3* meat processing plant; *subsystem 4* retail; *subsystem 5* household use

FU: 1 kg of beef meat; 1 kg of bone and fat free meat; 1 kg of beef meat (slaughter weight); 1 kg of beef meat product

GWP Global warming potential; *AP* acidification potential; *EP* eutrophication potential; *OLD* ozone layer depletion; *HT* human toxicity; *LC* land competition/use; *EC* energy consumption; *WC* water consumption; *FEP* fresh water aquatic ecotoxicity; *RD* resource depletion

4.3 Poultry Meat

It is considered that the poultry sector is the fastest-growing livestock sector as a result of the global dietary demand for healthy high-protein and low-fat type of meat (FAO 2013; OECD-FAO 2016). Globally, poultry is the most consumed meat after pork (13.8 compared to 15.3 kg/capita/year, respectively) (FAO 2015). Similar to other types of meat, EPIs in the poultry sector are water and energy consumption, feed production, wastewater discharge, and waste treatment (Bengtsson and Seddon 2013; González-García et al. 2014). The predominant environmental footprints related to the chicken meat chain are global warming potential, acidification potential, eutrophication potential, and cumulative energy demand, as well as ozone layer depletion (Skunca et al. 2018). Deeper analysis of the papers shows that Pardo's study is concentrated on potential improvements and not on in-depth LCA analysis of all five subsystems (Pardo et al. 2012) while others focus their research on farms and slaughterhouses. There is also a big diversity in functional units presented in these papers (live weight chicken, carcass weight, packaged broiler chicken, and broiler chicken products). Finally, different inventory was used in all studies raising concern on the comparability of the data.

Table 8 Summary of studies linking environmental impacts to poultry meat chain

Authors	Research focus	System boundaries					Environmental impact
		1	2	3	4	5	
Cesari et al. (2017)	LCA of broiler farm and slaughterhouse	✓	✓				GWP, AP, EP, CED
Pishgar-Komleh et al. (2017)	LCA of broiler chicken farms	✓					GWP, CED
Wiedemann et al. (2017)	LCA of chicken production	✓	✓	✓			GWP, CED
Kalhor et al. (2016)	LCA of broiler farm and slaughterhouse	✓	✓				GWP, AP, EP, OLD
González-García et al. (2014)	LCA of broiler chicken production	✓	✓				GWP, AP, EP, CED
Da Silva et al. (2014)	LCA of broiler chicken production system	✓	✓				GWP, AP, EP, CED
Thévenot et al. (2013)	LCA of poultry production	✓	✓				GWP, AP, EP, CED
Leinonen et al. (2012)	LCA of broiler production systems			✓			GWP, AP, EP, CED
Grandl et al. (2012)	Environment impacts and selected import sources	✓			✓		CED
Pardo et al. (2012)	Environmental improvement through LCA methodology	✓	✓	✓	✓	✓	GWP, AP, EP, CED

System boundaries: 1 chicken farm; 2 slaughterhouse; 3 meat processing plant; 4 retail; 5 household use

FU: 1 kg live weight chicken; 1 kg carcass weight; 1 kg packaged broiler chicken; 1 kg tray of sliced chicken breast packaged in modified atmosphere; 1 kg broiler chicken product

GWP global warming potential; AP acidification potential; EP eutrophication potential; CED cumulative energy demand; OLD ozone layer depletion

Table 8 present a short summary of manuscripts covering at LCA of least one part of the poultry meat chain.

Overall results show a large range of results for all environmental potentials. GWP ranges from below 0.25 kg CO_{2eq}/FU up to over 6.5 kg CO_{2eq}/FU depending on the subsystems observed, inventory and FU. AP reaches values up to 0.25 kg SO_{2eq} per kg of FU while EP goes from 0.002 to 0.085 kg PO_{4e}^{3-}.

The most examined subsystems are farms in line with the opinion that the highest impacts are on farms. However, Skunca et al. (2018) in their research covering more than 100 farms, slaughterhouses, meat processors, and retailers, as well as 500 households confirm that the average score of 1.81 kg CO_{2eq} was obtained at farms as in all other four subsystems together. This brings to attention the need to analyze all subsystems, namely retail and households since dietary and household habits influence environmental impacts among consumers. The differences are observed in terms of energy efficiency of refrigerators and freezers, different storage time of chicken meat, and chicken meat products in refrigerator and/or freezer and different cooking time of chicken meat and chicken meat products. GWP results were between 0.12 and 1.19 kg CO_{2eq}, CED results ranged between 1.77 and 23.2 MJ, while OLD results were between 0.32 and 318 μg CFC-11_{eq} (Skunca et al. 2018).

Farm activities have the highest environmental impacts in all footprints—GWP, AP, EP, and CED and crucial environmental hotspot for environmental impact categories is production of feed (Skunca et al. 2018).

5 Environmental Impact of the Meat Chain

Considering the environmental impact throughout the meat chain, Fig. 3 depicts the severity and timescale of environmental impacts on the five links in the meat chain from a functional unit point of view. The most severe and long-lasting environmental impact is at the farm stage. Slaughtering is an activity that lasts short (related to one animal) but the overall impact of slaughtering is high. Within retails, meat can be stored for a long period of time, but the environmental impact is not so high. Finally, the lowest impact is within meat processing where the meat processing activity (per FU) is short and at households where meat is often consumed within 7 days from purchasing. At both premises, environmental impact is not considered as high.

Sensitivity analysis is usually performed to distinguish between the influence and the importance of certain input parameters on the change of results. This type of analysis classifies parameters that identify potential mitigation strategies (Groen et al. 2016; Tassielli et al. 2018).

Figure 4 shows the influence and importance of parameters related to sensitivity analysis within the meat chain with four quadrants. The horizontal axis ranks the most influential parameters (ranked from low to high), and the vertical axis ranks the most important parameters (ranked from low to high). Reduction of influential

		Environmental impact	
		Low	High
Time-scale	Long	Retail	Farms
	Short	Meat processing Households	Slaughterhouses

Fig. 3 Severity and timescale of environmental impacts on the five links in the meat chain

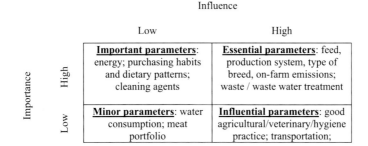

Fig. 4 Influence and importance of parameters related to sensitivity analysis in the meat chain

parameters may cause reduction of environmental impacts while important parameters reflect output uncertainty. Essential parameters are both influential and important, while minor parameters have low influence and importance (Skunca et al. 2018).

5.1 Environmental Management Systems in Meat Production

An environmental management system (EMS) is part of the management system used to manage environmental aspects, fulfill compliance and address risks and opportunities (ISO 2015). Most of implemented EMS worldwide are based on ISO 14001 (latest version from 2015) and an EMS is a part of an integrated management system (Labodová 2004). In the food industry (including the meat chain), EMS is usually integrated with quality management and/or food safety systems (Djekic et al. 2014). Besides food safety/quality dimensions, other standards/requirements often seen in meat production are standards covering animal welfare that measure conditions resulting from bad management practices, neglect, abuse of animals, or inadequately designed equipment (Grandin 2010) or requirements related to the religious component of slaughtering. Two global commercially accepted religious slaughtering methods are the "Halal" and "Kosher" methods of slaughtering practiced by Muslims and Jews respectively (Farouk 2013). Religious slaughtering in the EU is carried out in licensed slaughterhouses by authorized slaughter-men of the Jewish and Islamic faiths (Velarde et al. 2014).

As of the end of 2016, more than 340,000 EMS certificates were issued in over 200 countries, where the food chain participates with less than 3% (ISO 2017). A growing number of EMS certificates worldwide recognizes EMS as one of companies' priorities (Kimitaka 2010). However, there is no data regarding for the number of certificates in the meat chain worldwide (Djekic and Tomasevic 2016).

Companies interested in implementing an EMS expect to improve their environmental performance and enhance better company image (Massoud et al. 2010) or to enter international markets (Zeng et al. 2005). In order to develop an EMS and improve its environmental performance, a food organization has to assess its impacts and set environmental targets to reduce them (Djekic et al. 2014). Standard ISO 14001 promotes deployment of environmental impacts towards sustainable resource management and climate change mitigation including life cycle approach and effective communication with stakeholders (ISO 2015). Djekic et al. (2016) indicated significant differences in the levels of implementation of environmental practice with respect to the size of the meat companies, certification status, and meat sector—slaughterhouse or meat processing plant.

5.2   Route to Environmental Improvements

Environmental improvements in meat production have two opposed strategies, from changing dietary habits to specific improvement scenarios. Avoiding meat due to its environmental impact and/or animal welfare misses the goal due to the complexity of meat chain compared to other food chains (Röös et al. 2014). Swedish research on dietary changes in line with prevailing guidelines for a healthy meat intake confirmed that reduction of meat intake reduces GWP change, but variations in production systems and uncertainties in the calculation methodologies affect the results and conclusions much more (Hallström et al. 2014). Sustainable food industry should focus on pollution prevention, environmental, and technological improvements rather than discussing nutritional needs (Djekic and Tomasevic 2016).

In order to decrease the GWP and AP in meat production, focus should be on (1) manure management and (2) improving feeding strategy (Djekic et al. 2015). Gerber et al. (2015) suggest balancing feed ration and feed supplementation as well as animal health improvements at the farm stage. McAuliffe et al. (2016) believe that environmental impact of this developing technology in pig production will utilize manure as a source of biogas through anaerobic digestion. Also, manure management should be focused on improving on-site practices and/or manure quality (Djekic et al. 2015). Besides manure management, improvement of environmental management throughout the meat chain by fostering best environmental practices should be implemented (Djekic et al. 2016; Gerber et al. 2015). Also in line with (environmental) practices on-site, focus should be on the cold chain by decreasing the use of refrigerants with high GWP and developing new environmental and ozone-friendly refrigerants throughout the cold chain (Xu et al. 2015).

Finally, consumers in the meat chain are becoming more demanding in terms of diet requirements, food preservation technologies, and promotion of novel non-thermal technologies and food packaging, and these issues should also be considered in future analysis of environmental impact of the meat chain (Djekic et al. 2018).

6 Concluding Remarks

Analysis of the environmental impact of the meat chain is very complex and this food chain is considered as one of the food chains with global environmental impacts. Main challenges are due to different model assumptions and system boundaries when setting the LCA as well as various functional units in which environmental impacts are calculated making benchmarking throughout the meat chain difficult.

Regardless of the type of meat produced and technology applied, eating habits and cultural diversity, this type of production influences climate change in respect to global warming, acidification and eutrophication potentials and ozone depletion substances and has a high ratio of consumption of water and energy resulting in waste and wastewater discharge.

Three edges of the "environmental meat chain triangle" are the consumer, the environment, and the meat producers. The area within the triangle represents the improvement opportunity and potentials for future development in terms of consumers' dietary habits and sustainable meat production.

References

Allievi, F., Vinnari, M., & Luukkanen, J. (2015). Meat consumption and production—Analysis of efficiency, sufficiency and consistency of global trends. *Journal of Cleaner Production, 92,* 142–151.

Baldera Zubeldia, B., Nieto Jiménez, M., Valenzuela Claros, M. T., Mariscal Andrés, J. L., & Martin-Olmedo, P. (2016). Effectiveness of the cold chain control procedure in the retail sector in Southern Spain. *Food Control, 59,* 614–618.

Basset-Mens, C., & van der Werf, H. M. G. (2005). Scenario-based environmental assessment of farming systems: the case of pig production in France. *Agriculture, Ecosystems & Environment, 105*(1–2), 127–144.

Bengtsson, J., & Seddon, J. (2013). Cradle to retailer or quick service restaurant gate life cycle assessment of chicken products in Australia. *Journal of Cleaner Production, 41,* 291–300.

Bezirtzoglou, C., Dekas, K., & Charvalos, E. (2011). Climate changes, environment and infection: Facts, scenarios and growing awareness from the public health community within Europe. *Anaerobe, 17*(6), 337–340.

Borrisser-Pairó, F., Kallas, Z., Panella-Riera, N., Avena, M., Ibáñez, M., Olivares, A., et al. (2016). Towards entire male pigs in Europe: A perspective from the Spanish supply chain. *Research in Veterinary Science, 107,* 20–29.

Bragaglio, A., Napolitano, F., Pacelli, C., Pirlo, G., Sabia, E., Serrapica, F., et al. (2018). Environmental impacts of Italian beef production: A comparison between different systems. *Journal of Cleaner Production, 172,* 4033–4043.

Buratti, C., Fantozzi, F., Barbanera, M., Lascaro, E., Chiorri, M., & Cecchini, L. (2017). Carbon footprint of conventional and organic beef production systems: An Italian case study. *Science of the Total Environment, 576,* 129–137.

Carlsson-Kanyama, A. (1998). Climate change and dietary choices — how can emissions of greenhouse gases from food consumption be reduced? *Food Policy, 23*(3–4), 277–293

Carvalho, A., Mimoso, A. F., Mendes, A. N., & Matos, H. A. (2014). From a literature review to a framework for environmental process impact assessment index. *Journal of Cleaner Production, 64*, 36–62.

Cederberg, C., & Flysjö, A., (2004). *Environmental assessment of future pig farming systems— Quantifications of three scenarios from the FOOD 21 synthesis work*. The Swedish Institute for food and agriculture.

Cesari, V., Zucali, M., Sandrucci, A., Tamburini, A., Bava, L., & Toschi, I. (2017). Environmental impact assessment of an Italian vertically integrated broiler system through a life cycle approach. *Journal of Cleaner Production, 143*, 904–911.

Cherubini, E., Zanghelini, G. M., Alvarenga, R. A. F., Franco, D., & Soares, S.R. (2014). Life cycle assessment of swine production in Brazil: A comparison of four manure management systems. Journal of Cleaner Production (accepted manuscript).

Coulomb, D. (2008). Refrigeration and cold chain serving the global food industry and creating a better future: two key IIR challenges for improved health and environment. *Trends in Food Science & Technology, 19*(8), 413–417.

Čuček, L., Klemeš, J. J., & Kravanja, Z. (2015). Overview of environmental footprints. In *Assessing and measuring environmental impact and sustainability* (pp. 131–193). Elsevier.

Da Silva, V. P., van der Werf, H. M., Soares, S. R., & Corson, M. S. (2014). Environmental impacts of French and Brazilian broiler chicken production scenarios: An LCA approach. *Journal of Environmental Management, 133*, 222–231.

Dalgaard, R., Halberg, N., & Hermansen, J. E. (2007). Danish pork production—An environmental assessment. *DJF Animal Science* (University of Aarhus, Faculty of Agricultural Sciences).

De Backer, C. J., & Hudders, L. (2015). Meat morals: Relationship between meat consumption consumer attitudes towards human and animal welfare and moral behavior. *Meat Science, 99*, 68–74.

de Vries, M., & de Boer, I. J. M. (2010). Comparing environmental impacts for livestock products: A review of life cycle assessments. *Livestock Science, 128*(1–3), 1–11.

De Vries, M., Van Middelaar, C., & De Boer, I. (2015). Comparing environmental impacts of beef production systems: A review of life cycle assessments. *Livestock Science, 178*, 279–288.

Đekić, I. (2009). *Upravljanje zaštitom životne sredine u proizvodnji hrane* (1st ed.). Beograd, Srbija: Poljoprivredni fakultet Univerziteta u Beogradu.

Đekić, I., & Tomašević, I. (2017a). Environmental footprints in the meat chain. In *IOP Conference Series: Earth and Environmental Science* (Vol. 85(1), p. 012015).

Đekić, I., & Tomašević, I. (2017b). Environmental footprints in the meat chain. In *IOP Conference Series: Earth and Environmental Science* (p. 012015). IOP Publishing.

Delgado, C. L. (2003). Rising consumption of meat and milk in developing countries has created a new food revolution. *The Journal of Nutrition, 133*(11), 3907S–3910S.

Dick, M., Abreu da Silva, M., & Dewes, H. (2015). Life cycle assessment of beef cattle production in two typical grassland systems of Southern Brazil. *Journal of Cleaner Production, 96*, 426–434.

Djekic, I. (2015). Environmental impact of meat industry—Current status and future perspectives. *Procedia Food Science, 5*, 61–64.

Djekic, I., Blagojevic, B., Antic, D., Cegar, S., Tomasevic, I., & Smigic, N. (2016). Assessment of environmental practices in Serbian meat companies. *Journal of Cleaner Production, 112*(Part 4), 2495–2504.

Djekic, I., Radović, Č., Lukić, M., Stanišić, N., & Lilić, S. (2015). Environmental life-cycle assessment in production of pork products. *Meso, XVII*(5), 345–351.

Djekic, I., Rajkovic, A., Tomic, N., Smigic, N., & Radovanovic, R. (2014). Environmental management effects in certified Serbian food companies. *Journal of Cleaner Production, 76*, 196–199.

Djekic, I., Sanjuán, N., Clemente, G., Jambrak, A. R., Djukić-Vuković, A., Brodnjak, U. V., et al. (2018). Review on environmental models in the food chain—Current status and future perspectives. *Journal of Cleaner Production, 176*, 1012–1025.

Djekic, I., & Tomasevic, I. (2016). Environmental impacts of the meat chain—Current status and future perspectives. *Trends in Food Science & Technology, 54*, 94–102.

Domínguez-Rodrigo, M., Pickering, T. R., Diez-Martín, F., Mabulla, A., Musiba, C., Trancho, G., et al. (2012). Earliest porotic hyperostosis on a 1.5-million-year-old hominin, Olduvai Gorge, Tanzania. *PLoS ONE, 7*(10), e46414.

Dubey, R., Gunasekaran, A., & Samar Ali, S. (2015). Exploring the relationship between leadership, operational practices, institutional pressures and environmental perfo rmance: A framework for green supply chain. *International Journal of Production Economics, 160*, 120–132.

EC. (2009). Commission regulation (EC) No 1069/2009 of the European Parliament and the Council of 21 October 2009 laying down health rules as regards animal by-products and derived products not intended for human consumption and repealing Regulation (EC) No 1774/2002 (Animal by-products Regulation). In O. J. o. E. Union (Ed.). Official Journal of the European Union, Brussels, Belgium, pp. 1–95.

FAO. (2013). *Poultry development review*. Retrieved from http://www.fao.org/docrep/019/i3531e/i3531e.pdf. Accessed January 2018.

FAO. (2015). Retrieved from http://www.fao.org/docrep/005/y4252e/y4252e05b.htm.

Farouk, M. M. (2013). Advances in the industrial production of halal and kosher red meat. *Meat Science, 95*(4), 805–820.

Font-i-Furnols, M., & Guerrero, L. (2014). Consumer preference, behavior and perception about meat and meat products: An overview. *Meat Science, 98*(3), 361–371.

Gerber, P. J., Mottet, A., Opio, C. I., Falcucci, A., & Teillard, F. (2015). Environmental impacts of beef production: Review of challenges and perspectives for durability. *Meat Science, 109*, 2–12.

González-García, S., Gomez-Fernández, Z., Dias, A. C., Feijoo, G., Moreira, M. T., & Arroja, L. (2014). Life cycle assessment of broiler chicken production: A Portuguese case study. *Journal of Cleaner Production, 74*, 125–134.

Grandin, T. (2010). Auditing animal welfare at slaughter plants. *Meat Science, 86*(1), 56–65.

Grandl, F., Alig, M., Mieleitner, J., Nemecek, T., & Gaillard, G. (2012). Environmental impacts of different pork and chicken meat production systems in Switzerland and selected import sources. In 8th International Conference on LCA in the Agri-Food Sector (pp. 554–559).

Groen, E., Van Zanten, H., Heijungs, R., Bokkers, E., & De Boer, I. (2016). Sensitivity analysis of greenhouse gas emissions from a pork production chain. *Journal of Cleaner Production, 129*, 202–211.

Hallström, E., Röös, E., & Börjesson, P. (2014). Sustainable meat consumption: A quantitative analysis of nutritional intake, greenhouse gas emissions and land use from a Swedish perspective. *Food Policy, 47*, 81–90.

Hawkesworth, S., Dangour, A. D., Johnston, D., Lock, K., Poole, N., Rushton, J., et al. (2010). Feeding the world healthily: The challenge of measuring the effects of agriculture on health. *Philosophical Transactions of the Royal Society of London B: Biological Sciences, 365*(1554), 3083–3097.

Henchion, M., McCarthy, M., Resconi, V. C., & Troy, D. (2014). Meat consumption: Trends and quality matters. *Meat Science, 98*(3), 561–568.

Henri, J.-F., & Journeault, M. (2008). Environmental performance indicators: An empirical study of Canadian manufacturing firms. *Journal of Environmental Management, 87*(1), 165–176.

Herva, M., Franco, A., Carrasco, E. F., & Roca, E. (2011). Review of corporate environmental indicators. *Journal of Cleaner Production, 19*(15), 1687–1699.

Hoekstra, A. Y. (2008). Value of Water Research Report Series 28: Water neutral: Reducing and offsetting water footprints. Delft, The Netherlands: Unesco-IHE Institute for Water Education.

Holvoet, K., Sampers, I., Seynnaeve, M., & Uyttendaele, M. (2014). Relationships among hygiene indicators and enteric pathogens in irrigation water, soil and lettuce and the impact of climatic conditions on contamination in the lettuce primary production. *International Journal of Food Microbiology, 171*, 21–31.

Huerta, A. R., Güereca, L. P., & Lozano, Mdl S R. (2016). Environmental impact of beef production in Mexico through life cycle assessment. *Resources, Conservation and Recycling, 109,* 44–53.

IPCC. (2006). *IPCC guidelines for national greenhouse gas inventories intergovernmental panel on climate change.* Hayama, Kanagawa, Japan.

IPPC. (2006). *Integrated pollution prevention and control, reference document on best available techniques in the food, drink and milk industries.* Seville, Spain: European Commission.

ISO. (2006). *ISO 14040:2006 Environmental management—Life cycle assessment—Principles and framework.* Geneva, Switzerland: International Organization for Standardization.

ISO. (2015). *ISO 14001:2004 environmental management systems—Requirements with guidance for use.* Geneva, Switzerland: International Organization for Standardization.

ISO. (2017). *The ISO survey of certifications 2016.* Geneva, Switzerland: International Organization for Standardization.

Jasch, C. (2000). Environmental performance evaluation and indicators. *Journal of Cleaner Production, 8*(1), 79–88.

Kalhor, T., Rajabipour, A., Akram, A., & Sharifi, M. (2016). Environmental impact assessment of chicken meat production using life cycle assessment. *Information Processing in Agriculture, 3*(4), 262–271.

Kimitaka, N. (2010). Demand for ISO 14001 adoption in the global supply chain: An empirical analysis focusing on environmentally conscious markets. *Resource and Energy Economics, 32*(3), 395–407.

Kirezieva, K., Jacxsens, L., van Boekel, M. A. J. S., & Luning, P. A. (2015). Towards strategies to adapt to pressures on safety of fresh produce due to climate change. *Food Research International, 68,* 94–107.

Kratena, K. (2008). From ecological footprint to ecological rent: An economic indicator for resource constraints. *Ecological Economics, 64*(3), 507–516.

Kupusovic, T., Midzic, S., Silajdzic, I., & Bjelavac, J. (2007). Cleaner production measures in small-scale slaughterhouse industry—Case study in Bosnia and Herzegovina. *Journal of Cleaner Production, 15*(4), 378–383.

Labodová, A. (2004). Implementing integrated management systems using a risk analysis based approach. *Journal of Cleaner Production, 12*(6), 571–580.

Lal, A., Hales, S., French, N., & Baker, M. G. (2012). Seasonality in human zoonotic enteric diseases: A systematic review. *PLoS ONE, 7*(4), e31883.

Leinonen, I., Williams, A., Wiseman, J., Guy, J., & Kyriazakis, I. (2012). Predicting the environmental impacts of chicken systems in the United Kingdom through a life cycle assessment: Broiler production systems. *Poultry Science, 91*(1), 8–25.

Liu, C., Hofstra, N., & Franz, E. (2013). Impacts of climate change on the microbial safety of pre-harvest leafy green vegetables as indicated by *Escherichia coli* O157 and *Salmonella* spp. *International Journal of Food Microbiology, 163*(2–3), 119–128.

Liu, C., Hofstra, N., & Leemans, R. (2015). Preparing suitable climate scenario data to assess impacts on local food safety. *Food Research International, 68,* 31–40.

MacLeod, M., Gerber, P., Mottet, A., Tempio, G., Falcucci, A., Opio, C., et al. (2013). *Greenhouse gas emissions from pig and chicken supply chains—A global life cycle assessment.* Rome: Food and Agriculture Organization of the United Nations.

Massoud, M. A., Fayad, R., El-Fadel, M., & Kamleh, R. (2010). Drivers, barriers and incentives to implementing environmental management systems in the food industry: A case of Lebanon. *Journal of Cleaner Production, 18*(3), 200–209.

Mathijs, E. (2015). Exploring future patterns of meat consumption. *Meat Science, 109,* 112–116.

McAuliffe, G. A., Chapman, D. V., & Sage, C. L. (2016). A thematic review of life cycle assessment (LCA) applied to pig production. *Environmental Impact Assessment Review, 56,* 12–22.

Mekonnen, M. M., & Hoekstra, A. Y. (2010). *The green, blue and grey water footprint of farm animals and animal products.* Delft, The Netherlands: UNESCO-IHE Institute for Water Education.

Miraglia, M., Marvin, H. J. P., Kleter, G. A., Battilani, P., Brera, C., Coni, E., et al. (2009). Climate change and food safety: An emerging issue with special focus on Europe. *Food and Chemical Toxicology, 47*(5), 1009–1021.

Nguyen, T. L. T., Hermansen, J. E., & Mogensen, L. (2010). Environmental consequences of different beef production systems in the EU. *Journal of Cleaner Production, 18*(8), 756–766.

Nguyen, T. L. T., Hermansen, J. E., & Mogensen, L. (2011). *Environmental assessment of Danish pork*. Aarhus, Denmark: Aarhus University.

Nguyen, T. L. T., Hermansen, J. E., & Mogensen, L. (2012a). Environmental costs of meat production: The case of typical EU pork production. *Journal of Cleaner Production, 28,* 168–176.

Nguyen, T. T. H., van der Werf, H. M. G., Eugène, M., Veysset, P., Devun, J., Chesneau, G., et al. (2012b). Effects of type of ration and allocation methods on the environmental impacts of beef-production systems. *Livestock Science, 145*(1), 239–251.

Noya, I., Aldea, X., González-García, S. M., Gasol, C., Moreira, M. T., Amores, M. J., et al. (2017). Environmental assessment of the entire pork value chain in Catalonia—A strategy to work towards Circular Economy. *Science of the Total Environment, 589,* 122–129.

OECD/FAO. (2017a). *OECD-FAO agricultural outlook 2017–2026* (Special focus South East Asia). OECD Publishing.

OECD/FAO. (2017b). *OECD-FAO agricultural outlook 2017–2026* (Summary). OECD Publishing.

OECD-FAO. (2016). *OECD-FAO agricultural outlook 2016–2025*. Retrieved from http://www.fao.org/3/a-i5778e.pdf. Accessed January 2018.

Ogino, A., Sommart, K., Subepang, S., Mitsumori, M., Hayashi, K., Yamashita, T., et al. (2016). Environmental impacts of extensive and intensive beef production systems in Thailand evaluated by life cycle assessment. *Journal of Cleaner Production, 112,* 22–31.

Pardo, G., Ciruelos, A., Lopez, N., Gonzalez, L., Ramos, S., & Zufia, J. (2012). Environment improvement of a chicken product through life cycle assessment methodology. In 8th Conference on LCA in the Agri-Food Sector (pp. 86–91).

Parker, J. K., McIntyre, D., & Noble, R. T. (2010). Characterizing fecal contamination in stormwater runoff in coastal North Carolina, USA. *Water Research, 44*(14), 4186–4194.

Pennington, D. W., Potting, J., Finnveden, G., Lindeijer, E., Jolliet, O., Rydberg, T., et al. (2004). Life cycle assessment part 2: Current impact assessment practice. *Environment International, 30*(5), 721–739.

Pishgar-Komleh, S. H., Akram, A., Keyhani, A., & van Zelm, R. (2017). Life cycle energy use, costs, and greenhouse gas emission of broiler farms in different production systems in Iran—A case study of Alborz province. *Environmental Science and Pollution Research, 24*(19), 16041–16049.

Reckmann, K., Traulsen, I., & Krieter, J. (2012). Environmental impact assessment—Methodology with special emphasis on European pork production. *Journal of Environmental Management, 107,* 102–109.

Röös, E., Ekelund, L., & Tjärnemo, H. (2014). Communicating the environmental impact of meat production: challenges in the development of a Swedish meat guide. *Journal of Cleaner Production, 73,* 154–164.

Röös, E., Sundberg, C., Tidåker, P., Strid, I., & Hansson, P.-A. (2013). Can carbon footprint serve as an indicator of the environmental impact of meat production? *Ecological Indicators, 24,* 573–581.

Roy, P., Orikasa, T., Thammawong, M., Nakamura, N., Xu, Q., & Shiina, T. (2012). Life cycle of meats: An opportunity to abate the greenhouse gas emission from meat industry in Japan. *Journal of Environmental Management, 93*(1), 218–224.

Ruby, M. B. (2012). Vegetarianism. A blossoming field of study. *Appetite, 58*(1), 141–150.

Rahman, U. u., Sahar, A., & Khan, M. A. (2014). Recovery and utilization of effluents from meat processing industries. *Food Research International, 65*(Part C(0)), 322–328.

Sans, P., & Combris, P. (2015). World meat consumption patterns: An overview of the last fifty years (1961–2011). *Meat Science, 109,* 106–111.

Santos, G., Mendes, F., & Barbosa, J. (2011). Certification and integration of management systems: The experience of Portuguese small and medium enterprises. *Journal of Cleaner Production, 19*(17–18), 1965–1974.

Skunca, D., Tomasevic, I., Nastasijevic, I., Tomovic, V., & Djekic, I. (2018). Life cycle assessment of the chicken meat chain. *Journal of Cleaner Production, 184*, 440–450.

Sofos, J. N. (2014). Chapter 6—Meat and meat products. In Y. M. Lelieveld (Ed.), *Food safety management* (pp. 119–162). San Diego: Academic Press.

Speth, J. D. (1989). Early hominid hunting and scavenging: The role of meat as an energy source. *Journal of Human Evolution, 18*(4), 329–343.

Tassielli, G., Notarnicola, B., Renzulli, P., & Arcese, G. (2018). Environmental life cycle assessment of fresh and processed sweet cherries in southern Italy. *Journal of Cleaner Production, 171*, 184–197.

Thévenot, A., Aubin, J., Tillard, E., & Vayssières, J. (2013). Accounting for farm diversity in life cycle assessment studies—The case of poultry production in a tropical island. *Journal of Cleaner Production, 57*, 280–292.

Tsutsumi, M., Ono, Y., Ogasawara, H., & Hojito, M. (2018). Life-cycle impact assessment of organic and non-organic grass-fed beef production in Japan. *Journal of Cleaner Production, 172*, 2513–2520.

UNEP. (2000). Cleaner production assessment in meat processing. In: D.e.p.a.-D.M.o.e.a. energy (Ed.). United Nations Environment Programme Division of Technology, Industry and Economics, Paris, France.

Uyttendaele, M., Liu, C., & Hofstra, N. (2015). Special issue on the impacts of climate change on food safety. *Food Research International, 68*, 1–6.

Velarde, A., Rodriguez, P., Dalmau, A., Fuentes, C., Llonch, P., von Holleben, K. V., et al. (2014). Religious slaughter: Evaluation of current practices in selected countries. *Meat Science, 96*(1), 278–287.

Wiedemann, S., McGahan, E., & Murphy, C. (2017). Resource use and environmental impacts from Australian chicken meat production. *Journal of Cleaner Production, 140*, 675–684.

Williams, A. G., Audsley, E., & Sandars, D. L. (2006). *Determining the environmental burdens and resource use in the production of agricultural and horticultural commodities*. Main Report. Defra Research Project IS0205. Bedford: Cranfield University and Defra.

WMO. (2017). *Global atmosphere watch program—greenhouse gas research*. Geneva, Switzerland: World Meteorological Organization, Atmospheric Environment Research Division, Research Department.

Xu, Z., Sun, D.-W., Zhang, Z., & Zhu, Z. (2015). Research developments in methods to reduce carbon footprint of cooking operations: A review. *Trends in Food Science & Technology, 44*(1), 49–57.

Zeng, S. X., Tam, C. M., Tam, V. W. Y., & Deng, Z. M. (2005). Towards implementation of ISO 14001 environmental management systems in selected industries in China. *Journal of Cleaner Production, 13*(7), 645–656.

Printed in the United States
By Bookmasters